Nuclear Materials
An Overview

Haydee Domenech

Retired Radiation Safety Expert
Miami, Florida

CRC Press
Taylor & Francis Group
Boca Raton London New York

CRC Press is an imprint of the
Taylor & Francis Group, an **informa** business

A SCIENCE PUBLISHERS BOOK

Cover credit: The image taken from Chapter 3 (Figure 3.16).

- **Left-hand side image:** A public domain photo of a PWR nuclear fuel assembly courtesy of the French Alternative Energies and Atomic Energy Commission.
- **Right-hand side image:** A public domain photo of CANDU fuel assemblies, and a partial view of a CANDU calandria that showed a fuel assembly, taken from a video from the Canadian Nuclear Safety Commission (a government agency).

First edition published 2025
by CRC Press
2385 NW Executive Center Drive, Suite 320, Boca Raton FL 33431

and by CRC Press
4 Park Square, Milton Park, Abingdon, Oxon, OX14 4RN

CRC Press is an imprint of Taylor & Francis Group, LLC

Library of Congress Cataloging-in-Publication Data (applied for)

ISBN: 978-1-032-35368-5 (hbk)
ISBN: 978-1-032-35370-8 (pbk)
ISBN: 978-1-003-32655-7 (ebk)

DOI: 10.1201/9781003326557

Typeset in Palatino Linotype
by Prime Publishing Services

Preface

In the face of climate change, the quest for sustainable energy sources is more urgent than ever. This book is intended to explore the domain of nuclear materials, a pivotal component in reducing greenhouse gas emissions. It gives an overview of the origins, uses, and impacts of nuclear materials on our world. Beginning with an introduction to the phenomenon of radioactivity and radioactive materials, the book records the prominent milestones related to the discovery of radioactive elements, subatomic particles, and nuclear reactions in the years preceding the Second World War, and their role in developing the atomic bomb.

The second chapter covers the history of post-war reactors and electricity production, and the creation of the International Atomic Energy Agency (IAEA) to promote the peaceful use of nuclear energy, emphasizing the commitment of the USA and the international community to preventing nuclear weapons proliferation worldwide. Here, it defines the concept of nuclear materials and discusses the importance of the Non-Proliferation Treaty (NPT), and its associated measures of safeguards, inspections, and the control and accountability of nuclear materials.

Next, the book provides an overview of the technologies to convert nuclear materials into nuclear fuel and the nuclear fuel cycle, from uranium mining and milling to fuel burnup during power generation. It also discusses the options for spent fuel interim storage, the alternatives for spent fuel reprocessing, and the advantages of deconversion, and re-enrichment of depleted uranium tails.

Safety is a critical aspect of working with nuclear materials, so the book also dedicates attention to this topic. First, it provides an overview of the importance of 'safety by design' of nuclear reactors and outlines the authorization process required for nuclear fuel cycle facilities. The book discusses radiation safety, safety analysis, nuclear criticality and safety, potential chemical hazards associated with uranium conversion, and the interface between security and safety in the nuclear fuel cycle.

Finally, while focusing on the safety and low environmental impact of nuclear energy, it examines the topic of radioactive waste resulting from the various stages of the nuclear fuel cycle, the current options for their management, and the options for the disposal of intermediate and high-level radioactive waste, alongside the low-level waste resulting from environmental remediation of past practices in uranium mining and the decommissioning of old reactors. It also formulates some thoughts for a range of future waste from advanced reactors thinking of actual and future generations.

Contents

Preface   iii

1. Radioactivity and Radioactive Materials   1
Naturally occurring radioactive materials   3
Nuclear reactions   6
Radioactive series   10
Neutron discovery and artificially induced radioactivity   11
Self-sustained chain reaction. The path to the release of   12
huge amounts of energy
References   15

2. Atoms for Peace   18
First post-war nuclear reactors and electricity production   18
Creation of the IAEA to promote the peaceful use of   22
nuclear energy
Nuclear material definition: source material and   25
special nuclear material
The Non-Proliferation Treaty, safeguards, and inspection   25
Control and accountability of nuclear material   29
Safeguards inspections and surveillance   34
References   38

3. Nuclear Fuel Cycle Overview   42
Uranium mining and milling   43
Uranium conversion   52
Uranium enrichment   55
Fuel fabrication   60
Fuel burnup during power generation   64
Spent fuel interim storage   64
Spent fuel reprocessing   67
Deconversion and re-enrichment   70
References   74

4. Safety in Nuclear Fuel Cycle Facilities   **81**
Authorization process   82
Radiation safety outline   82
Safety analysis   89
Outline of chemical hazards   89
Nuclear criticality safety outline   90
The interface between security and safety in the nuclear   92
fuel cycle
References   94

5. Overview of Nuclear Reactors   **96**
Some statistics on nuclear reactors   98
Nuclear reactor types and their components   99
Some milestones in advanced nuclear reactors   108
Advanced nuclear fuel development   112
Generation IV nuclear reactors   119
References   135

6. Nuclear Reactor Safety   **142**
Three areas of safety measures: radiation safety, nuclear   142
safety, and security
Safety systems, structures, and components   147
Defense-in-depth safety concept   149
Redundancy, diversity, and independence principles   151
Containment safety concept   152
Emergency planning and response   153
Safety culture   156
Safety in advanced reactors   157
References   159

7. Environmental Impact of Nuclear Materials   **164**
Radioactive waste from mining, milling, and processing   166
Radioactive waste from fuel fabrication and reprocessing plants   170
Radioactive waste from nuclear reactors   175
Spent nuclear fuel and high-level radioactive   182
waste management
Decommissioning of nuclear facilities   184
Consideration of future waste from advanced reactors   187
References   190

Index   **197**

CHAPTER 1

Radioactivity and Radioactive Materials

In the beginning of the 20th century, it was established that all matter is made up of exceedingly small and invisible units named atoms, which are, in turn, composed of a dense center identified as the nucleus and several smaller negative electrically charged particles known as electrons, orbiting the nucleus in specific shells restricted to certain discrete values of energy. The nucleus, within which the bulk of the atom mass is concentrated, is now understood to be a quantum system composed of protons and neutrons, particles of nearly equal mass and the same intrinsic angular momentum (spin) of ½, both made of quarks and held together by the strong force of gluons.

Atoms in nature are normally electrically neutral, meaning that the negative charges of the orbiting electrons are compensated by the same number of positive charges of protons in the nucleus. A nucleus of a certain atomic number, Z, and a specific mass number, A, ($^{A}_{Z}X$) is known as a nuclide. Nuclei are either stable or unstable. Nuclear stability depends on the proton-neutron ratio and the mass number. Most stable nuclei (up to about atomic number 20) have a neutron-to-proton ratio equal to 1. These nuclei have enough binding energy to hold the elements of the nucleus together. As the number of protons increases, the ratio of neutrons to protons needed to ensure a stable nucleus increases steadily to about 1.5. Those nuclides with atomic numbers greater than 83, the so-called heavy nuclei, have an excess of energy, making them unstable. Unstable nuclei are also known as radioactive nuclides, radioactive isotopes, or radionuclides. They tend to reach stability by emitting particles and electromagnetic radiation. The heaviest nuclei, such as those of the element uranium, tend toward stability by ejecting part of their mass, by converting to energy. For example, ^{238}U breaks forming a lighter nucleus

[234]Th and emits an alpha particle—which is essentially the helium nucleus. It has 2 protons and 2 neutrons, with a mass number of 4.

$$\ce{^{238}_{92}U} \rightarrow \ce{^{234}_{90}Th} + \ce{^{4}_{2}He}$$

As illustrated in Figure 1.1, radionuclides undergo spontaneous nuclear transformations by either decreasing their mass number, ejecting two protons and two neutrons (alpha decay), or by transforming protons and neutrons into each other within the nucleus to emit beta radiation in the form of an electron or a positron[1] (beta decay), which alters the structure but keeps the mass number same. Gamma radiation occurs when the nucleus changes from a higher-level energy state to a lower one, which often goes with spontaneous alpha and beta decay.

The process of spontaneous decay by which a nucleus loses energy emitting particles or electromagnetic radiation (photons) or transforms itself into another nucleus with a different atomic number (Z) is known as radioactivity. Emitted radiation energy is characteristic of each radionuclide. The type of emitted particle, its energy, and the time average of the emission will depend only on the nature of the radionuclide and not be altered by surrounding influences like pressure, temperature, electric or electromagnetic fields, or chemical reactions. All materials having

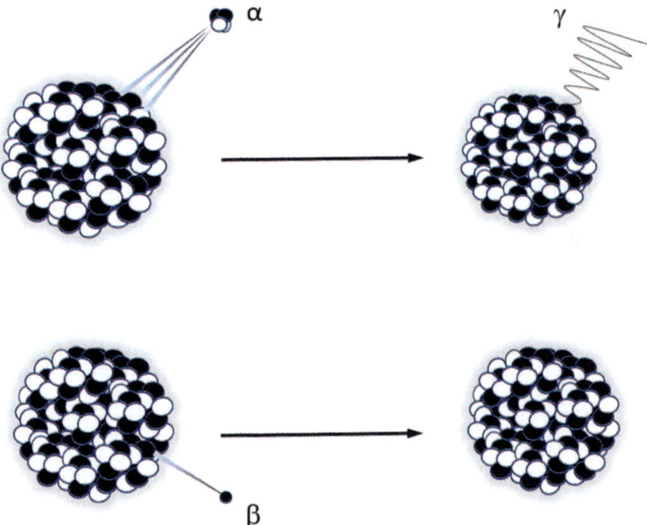

Figure 1.1. Radioactive decay by emitting α particles (mass number changes) or β particles.

---

[1] It is a particle identical to an electron except that it has a positive electrical charge.

unstable nuclei are known as radioactive materials. Radioactivity can be natural when it originates from radionuclides present in stars billions of years ago and cosmic radiation, and artificial when it is produced by bombarding stable nuclei with high-speed particles.

Massive nuclei with a high neutron-to-proton ratio can also undergo spontaneous fission, where they will split into two equivalent nuclei, or lighter elements, release free neutrons and photons, and large amounts of energy.

The physical quantity half-life (T½) is the time required for the nuclei population to decrease by a radioactive decay process by half. Activity, the quantity related to radioactivity, which is the number of disintegrations of a radionuclide per unit of time, also decreases by half by the same radioactive decay process, so the time taken for the activity of a radionuclide to decrease by half can be used as an alternative definition of half-life.

Naturally occurring radioactive materials

Natural resources mined from the ground, such as coal, oil, natural gas, and other mineral ores, include different amounts of radioactive elements of natural origin. According to the United States Environmental Protection Agency (EPA), Naturally Occurring Radioactive Materials (NORM) are "materials which may contain any of the primordial[2] radionuclides or radioactive elements as they occur in nature, such as uranium, thorium, potassium, and their radioactive decay products,[3] such as radium and radon, which are undisturbed as a result of human activities" [1]. Uranium and thorium in any form are also regarded as nuclear materials [2].

Uranium is the heaviest element found in large quantities on Earth. It is present in low concentrations in soil, rock, and water, and is commercially extracted from minerals such as uraninite (uranium dioxide, UO_2), pitchblende (triuranium octaoxide, U_3O_8), carnotite (hydrated potassium uranyl vanadate, $K_2(UO_2)_2(VO_4)_2.3H_2O$), autunite (hydrated calcium uranyl phosphate, $Ca(UO_2)_2(PO4)_2.10H_2O$), and torbernite (hydrated copper uranyl phosphate, $Cu(UO_2)_2(PO_4)_2.10H_2O$).

The element uranium was discovered in 1789 by Martin Klaproth, a German chemist, who isolated UO_2 while analyzing samples from the silver mines of Jáchymov in the Karlovy Vary Region of the Czech Republic. He named the element uranium after the planet Uranus.

---

[2] Primordial means existing from the beginning of time, naturally occurring.

[3] The atoms formed, and the energy and particles emitted as the radioactive material decay to reach a stable form.

All uranium isotopes[4] are radioactive. Natural uranium consists mostly of ^{238}U (T½ 4.5 billion years), less than 1% of ^{235}U (T½ 713,000,000 years), and trace amounts of ^{234}U (T½ 247,000 years). ^{235}U is the only naturally occurring fissile material. This means that it can undergo a spontaneous fission reaction.

Uranium is primarily used as fuel for nuclear reactors to produce electricity and heat, and run naval ships, submarines, and spaceships. Likewise, for reactors to produce isotopes with applications in medicine, research, food processing, and several other industries. Highly enriched uranium (^{235}U) is used in nuclear weapons.

Because of its high density and weight, uranium is also employed as a target material for the production of X-rays; as a component of shielding materials; in inertial guidance systems; in gyro compasses; as ballast for missile reentry vehicles; as a counterweight for aircraft control surfaces, and so on. Furthermore, uranium dating is valuable to date rocks that formed and crystallized about 4.5 billion years ago.

Thorium is primarily found in soil, rocks, water, plants, and animals in the form of ^{232}Th with a half-life of ~ 14 billion years—almost three times the age of the Earth. Thorium was identified in 1828 by the Swedish chemist Jöns Jakob Berzelius, who named it after Thor, the Norse god of thunder. Thorium isotopes ^{234}Th (T½ 24.1 days), ^{230}Th (T½ 75,200 years), and ^{228}Th (T½ 1.91 years), which are products of the decay chain of natural thorium (^{232}Th), can also be found naturally at trace levels, but their presence in terms of mass is negligible.

Thorium is obtained from monazite sands as a by-product of mining rare-earth metals like cerium, lanthanum, and neodymium, which are essential elements in making high-technology devices, including smartphones digital cameras, computer hard disks, etc. The generic formula of monazite is $(Ce,La,Nd,Th)(PO_4,SiO_4)$.

Thorium dioxide (ThO_2) is used to make camera and telescope lenses, ceramics, and fire bricks. Also, to increase the heat resistance of metal alloys. For example, it is used to reinforce magnesium alloys. In gas tungsten arc welding ThO_2 is used to increase the high-temperature strength of tungsten electrodes and improve arc stability [3]. The alloy of platinum and thorium is an effective catalyst for oxidizing ammonia to nitrogen oxides.

^{232}Th is not fissile and so is not good as fuel in thermal neutron reactors. However, it is fertile[5] and upon capturing neutrons it will transmute into

---

[4] Nuclides from the same element having the same atomic number and different mass number.
[5] Fertile means that it can be converted into a fissile material by irradiation in a reactor.

fissile [239]Pu and [233]U. The best fissile isotope for thermal neutrons is [233]U, which can be used for breeding in both thermal and fast reactors [4].

Potassium shares the natural radioactivity on Earth along with uranium and thorium. [40]K, with a T½ of 1.25 billion years, is found in the oceans, mineral waters, brine, soil, various minerals, etc. Even so, it represents only a very small fraction (about 0.012%) of naturally occurring potassium. [40]K is largely incorporated from the soil into the food we eat and into our body tissues, but the health risk from bodily [40]K to humans is as low as that of cosmic radiation.

The half-life of other potassium radionuclides that exist in addition to [40]K is negligible. [40]K decays into two different nuclei: [40]Ca by beta-negative decay, and [40]Ar by electron capture followed by gamma emission at an energy of 1.46 MeV. The argon isotope is particularly useful for dating rocks whose age is between a million and a billion years.

Radium and radon are formed by the decay of uranium and thorium, and all of their isotopes are radioactive. The most significant radium radionuclides are [226]Ra (T½ 1600 years), [228]Ra (T½ 5.75 years), [223]Ra (T½ 11.43 days), and [224]Ra (T½ 3.6 days). In addition to the primary radiation—alpha, beta, or both—most of the decay products of radium and its radionuclides emit other radiation such as X-rays, gamma rays, internal conversion electrons, and Auger electrons.[6]

Radon is a gas resulting from the decay of radium. Three radioactive isotopes of radon are found naturally in significant quantities: [222]Rn (T½ 3.8 days), [220]Rn (55.6 seconds), and [219]Rn (T½ 3.96 seconds). They are the result of the decay of the radionuclides present in the ground and building materials. Radon short-lived isotopes readily attach to dust particles, which are then inhaled into the lungs and deposited on the mucous lining of the respiratory tract. Long-term radon exposure in enclosed spaces can damage the airways and increase the risk of lung cancer.

Radium, along with polonium, was discovered in 1898 by French physicists Pierre and Marie Curie while researching the uranium ore pitchblende. After isolating radium, they shared the 1903 Nobel Prize in physics with French scientist A. Henri Becquerel for their studies on radioactivity.

Since its discovery, [226]Ra found several applications in cancer therapy, e.g., applicators for skin cancer treatment; intracavitary brachytherapy[7] for the treatment of cervix or uterus cancer; and interstitial brachytherapy needles for the treatment of prostate cancer and other superficial tumors.

---

[6] Very low energy electrons emitted by radionuclides that decay by electron capture.

[7] A procedure that involves placing hermetically sealed radioactive material inside the patient to kill cancer cells and shrink tumors.

Radium-beryllium neutron sources were also suitable for research works, and geophysical prospecting for oil. A luminescent paint made with ^{226}Ra was widely used for watch and clock dials, aircraft instruments, and some household items until the 1970s.

Adverse radiological effects of ^{226}Ra are attributable to its intake. In the late 1920s, it was established that it could lead to bone sarcomas. Exposure to its decay progeny, in particular, ^{222}Rn, inhalation of which is recognized as having potential to cause lung cancer [5]. It is also well known that sealed ^{226}Ra sources have been a major radiation protection problem over the years on account of the internal pressure generated by decay, which facilitates the leaking and spread of contamination. Consequently, when radionuclides like ^{60}Co, ^{90}Sr, ^{137}Cs, and ^{192}Ir produced artificially in nuclear reactors and accelerators became available, the production of radium was stopped, and radium's therapeutic applications were superseded by less costly and safer sealed sources [6].

Nuclear reactions

Nuclear decay and nuclear transmutation are both nuclear reactions. As stated before, a nuclear decay reaction (also called radioactive decay or radioactivity) is when an unstable nucleus transforms into nuclei of one or more other elements, emitting alpha particles, beta particles, or photons (γ radiation or X-rays). The resulting daughter[8] nuclei have a lower mass and are lower in energy, i.e., are more stable than the parent[9] nucleus. Nuclear decay reactions are spontaneous and are characteristic of each radionuclide. Following are examples of radioactive decay:

$$^{226}_{88}Ra \rightarrow {}^{222}_{86}Rn + {}^{4}_{2}\alpha$$

where ^{226}Ra emits an alpha particle to form ^{222}Rn

$$^{14}_{6}C \rightarrow {}^{11}_{5}B + {}^{0}_{+1}\beta^{+1}$$

where ^{14}C (carbon-14) undergoes positron emission to form ^{11}B (boron-11)

$$^{14}_{6}C \rightarrow {}^{14}_{7}N + {}^{0}_{1}\beta$$

where ^{14}C (carbon-14) undergoes beta (β) decay to form ^{14}N (nitrogen-14)

$$^{99m}_{43}Tc \rightarrow {}^{99}_{43}Tc + {}^{0}_{0}\gamma$$

where ^{99m}Tc (technetium-99 metastable, i.e., excited state) emits high energy photons (γ rays) to return to the ground state during its decay.

---

[8] The nucleus remaining after the decay is usually referred to as the daughter nucleus.
[9] The disintegrating nucleus is usually referred to as the parent nucleus.

The heaviest nuclei with an atomic mass ≥ 230, like natural uranium, can undergo spontaneous fission, where a nucleus is split into two smaller nuclei releasing energy and one or more neutrons. Spontaneous fission is a form of nuclear decay, although its probability is extremely low for naturally occurring radionuclides because of the long half-times. For example, ^{238}U with a half-time of 4.5×10^9 years, decays by spontaneous fission with a probability of 0.000055%. Yet, there is evidence of damage in the crystal structure of uranium-containing minerals owing to spontaneous fission.

Heavy transuranium elements, most of them artificial, can also go through spontaneous fission and release neutrons. A number of those radionuclides with some noteworthy probability can be used as neutron sources in several applications. For example, the spontaneous fission of ^{252}Cf (californium 252) occurs with a probability of 3.09%.

Spontaneous fission happens without externally added excitation energy, i.e., without the nucleus being collided with a neutron or other particles. For spontaneous fission to go on, a quantum mechanical process involving penetration of the potential barrier should occur [7, 8]. This process is also called quantum mechanical tunneling and explains how subatomic particles with less energy than the potential energy barrier can cross through the barrier [9] and generate fission.

Fission can also be induced artificially by exciting the nucleus to a particular value of energy for a certain nuclide. As illustrated in Figure 1.2, after a nucleus collides with a neutron carrying enough energy, it splits into two nuclei, called fission products—which together would be less massive than the original nucleus. The minimum excitation energy required for fission to occur is the critical energy or threshold energy.

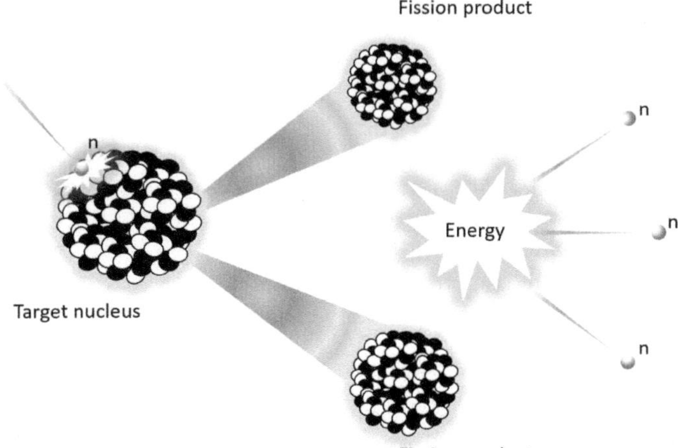

Figure 1.2. Nuclear fission process.

Consistent with Einstein's theory connecting small amounts of mass (m) with enormous amounts of energy (E): $E = m \times c^2$, fission would be accompanied by the release of an enormous amount of energy and the emission of several neutrons. Released neutrons can similarly induce fission in other neighboring nuclei of fissionable material, causing a chain reaction. If the chain reaction is controlled in a nuclear reactor, it can provide enough power to produce heat or electricity. If uncontrolled, it can lead to an explosion of tremendous destructive force.

Heavy nuclei can be classified as either fissile or fissionable materials, depending on the speed of the neutrons that cause fission. Fissile material is a nuclide that is capable of undergoing fission after absorbing a thermal neutron, e.g., ^{233}U, ^{235}U, ^{239}Pu. Fissionable material is a nuclide that is capable of fission after capturing either fast neutrons or thermal neutrons. This category includes nuclei that can only undergo fission with high-energy neutrons, e.g., ^{232}Th, ^{238}U [10].

Nuclear transmutation is a reaction where a nucleus is converted into another more massive nucleus when colliding with highly energetic particles, e.g., alpha particles, deuterons, small nuclei, neutrons, etc. Transmutation reactions occur only under special conditions, for instance, using particle accelerators or in the interior of stars.

The first successful nuclear transmutation reaction was carried out in 1919 by New Zealand physicist Ernest Rutherford, who showed that α particles emitted by radium could react with nitrogen nuclei to form oxygen nuclei as described below. This reaction also proved the existence of the proton as part of the nucleus.

$$\tfrac{4}{2}\alpha + \tfrac{14}{7}N \rightarrow \tfrac{17}{8}O + \tfrac{1}{1}p$$

Nuclear transmutation reactions can also be natural or artificial. Natural transmutation occurs in unstable, radioactive elements, which finally transform into a stable element over a decay series. For example, ^{238}U and ^{232}Th transmute spontaneously into ^{206}Pb and ^{208}Pb respectively through a sequence of steps. The same occurs in the core of stars, usually through fusion reactions between hydrogen isotopes and helium ($^{4}_{2}He$) but also between heavier elements up to iron.

Artificial transmutation needs potent accelerators at nearly the speed of light to take place. Heavy elements following uranium have been synthesized through transmutation. For example, neptunium, curium, einsteinium, and rutherfordium [11]. Transmutation processes have been proposed as well for the treatment of radioactive waste containing fission products [12, 13].

Fusion is a nuclear reaction that happens readily on the surface of the sun, where the temperatures are well more than those needed for overcoming the repulsive force between colliding hydrogen nuclei. The reaction of deuterium ($_1^2H$) and tritium ($_1^3H$) is:

$$_1^2H + {}_1^3H \rightarrow {}_2^4He + {}_0^1n + E$$

In a fusion reaction, two light nuclei merge to form a single heavier nucleus, as illustrated in Figure 1.3. The process releases energy because the mass of the resulting helium nucleus is less than the mass of the two original nuclei [14]. A fusion reaction is not a chain reaction, therefore, for each deuterium-tritium collision, only a single burst of energy is released.

Sustaining controlled fusion has still not been achievable after more than 50 years of research. However, key fusion plasma performance parameters have been increased by a factor of 10,000 over this time. Main efforts have been centered on tokamak reactors and stellarators which confine a deuterium-tritium plasma magnetically. A newer line of research is inertial confinement fusion, where laser or ion beams are focused very precisely onto the surface of the target[10] [15]. The U.S. National Ignition Facility (NIF), with the world's most energetic laser-based inertial confinement fusion research device, recently reported the achievement of a burning plasma state[11] [16].

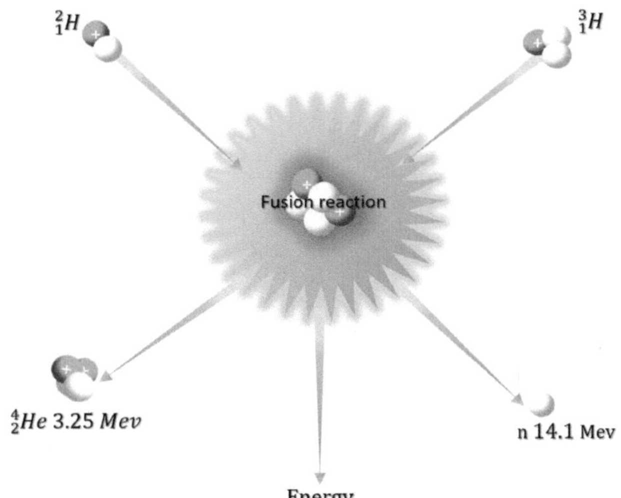

Figure 1.3. Nuclear fusion reaction of light elements.

---

[10] A small pellet few millimetres in diameter of deuterium-tritium.
[11] Where most of the plasma heating comes from fusion reactions.

Radioactive series

As mentioned before, the natural transmutation of unstable heavy nuclides (Z > 83) occurs through a sequence of steps until a stable nucleus is achieved. This sequence involves a succession of α and β decays taking place from parents to daughters. Given that α decay decreases Z by 2, and β decay[12] decreases Z by 1, it is almost impossible for any nuclide with Z > 83 to decay to a stable daughter nuclide in a single step, except by a nuclear fission reaction. Heavy nuclides decay to a different daughter nucleus that is radioactive, which in turn decays to another radioactive daughter nucleus, and so on. The sequence of transmutations to reach this balance is called the decay chain.

Radioactive series are three naturally occurring decay chains headed by primordial radionuclides. Given that they include radioactive elements accumulated over millions of years, many of the radioactive daughters are in equilibrium with uranium and thorium, also contributing to individual transformations.

As seen in Figure 1.4, the three naturally decaying series are the thorium series, beginning with natural ^{232}Th; the uranium series, beginning with natural ^{238}U—also known as the radium series; and the actinium series, beginning with naturally occurring ^{235}U.

Some decay reactions in the chains are accompanied by the emission of γ radiation, but they are not shown because such emission does not

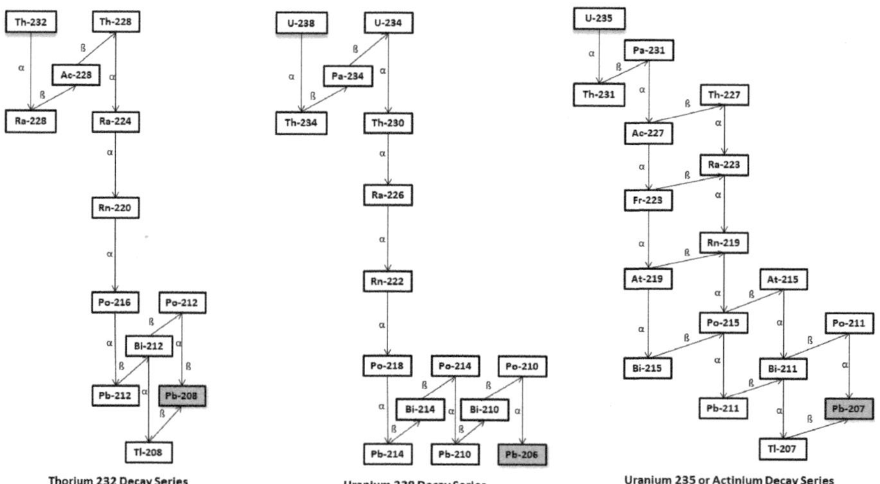

Figure 1.4. Naturally occurring radioactive series.

---

[12] Electron emission, positron emission or electron capture.

produce a daughter that differs from the parent. All three series end in stable isotopes of lead, i.e., ^{208}Pb, ^{207}Pb, and ^{206}Pb, appearing in gray in Figure 1.4.

There is a fourth series, the decay of ^{237}Np (neptunium-237) to ^{209}Bi (bismuth-209) and, finally, to ^{205}Tl (thallium-205) as a stable isotope. Neptunium decay series is known to have occurred in the remote past and to have decayed very long ago because the half-life of the longest-lived nuclide is short compared to the age of the earth. Only two of the nuclides concerned with this series are found naturally, ^{209}Bi and ^{205}Tl.

^{209}Bi undergoes α decay with a T½ of approximately 2.01×10^{19} years (20 quintillion years) to decay to ^{205}Tl, which means over a billion times longer than the current estimated age of the universe. To all effects, ^{209}Bi can be treated as if it were non-radioactive.

Neutron discovery and artificially induced radioactivity

Two of the most notable scientific breakthroughs took place in the 1930–1940 period. They were the discovery of the neutron and the discovery of artificially induced radioactivity. These discoveries preceded the fast developments in nuclear physics that went on before and after WWII.

Earlier significant outcomes were the first artificially induced nuclear reaction performed by Ernest Rutherford and his discovery of the proton; advancements that gave birth to the nuclear model settling on the foundations of nuclear physics [17]. Devices like the linear particle accelerator (Linac), the betatron, and the cyclotron—conceived and built between 1924 and 1932—were the driving force behind the study of the fundamental structure of matter at this time [18].

The discovery of the neutron by British physicist James Chadwick in 1932 [19] which occurred later, decisively established the essential character of the atomic nucleus. The newly discovered elementary particle, uncharged, with about the same mass as the proton, but dissimilar, meant a new direction in the study of nuclear structure. Shortly after finding the neutron, two new fundamental interactions were discovered, the strong interaction that binds nucleons together in the nucleus, and the weak interaction responsible for β decay [20].

The next year, in 1933, French chemist Irene Joliot-Curie (Marie Curie's daughter) and French physicist Frédéric Joliot, her husband, artificially synthesized a radioactive isotope for the first time [21]. They obtained ^{30}P (phosphorus-30) by bombarding aluminum with alpha particles. Their discovery anticipated the future production of radionuclides using highly energetic particles. What is more, it opened the door to the discovery of nuclear fission.

Following both crucial discoveries and in the search for transuranic elements, Italian physicist Enrico Fermi undertook an extensive study of the nuclear reactions that can be produced by the bombardment of various elements with neutrons. He believed neutrons might be more effective at creating artificial radioactivity than α particles because they would not be repelled by the positively charged nuclei of the bombarded atoms. He found that the best results were obtained by slowing the neutrons down before they hit their target atoms.

Pursuing Fermi's work, Austrian physicist Lise Meitner and German chemists Otto Hahn, and Fritz Strassmann [22] also conducted experiments using the bombardment of uranium and other elements with slow neutrons. As a result, in December of 1938, Meitner and Fritz concluded that uranium breaks into atoms of two lighter elements: barium, and krypton, uncovering that the mass excess from the original uranium was converted into an enormous amount of energy.

A short time later, in 1939, Frédéric Joliot and his colleagues [23] demonstrated the production of several neutrons in the fission of ^{235}U and the possibility of a self-sustaining chain reaction with a significant release of energy. The same year, Danish physicist Niels Bohr [24] suggested that ^{235}U is solely responsible for slow-neutron fission, while the abundant ^{238}U requires fast neutrons of over 1 MeV in energy to fission.

Self-sustained chain reaction. The path to the release of huge amounts of energy

The discovery of nuclear fission and chain reactions led to the culmination of a decisive chapter in the history of nuclear science, and the beginning of a new era. The friendly competition mentioned before between a group of brilliant scientists ended abruptly with the imminence of war. Even sooner than WWII formally began, many of them became suspicious and were intimidated by the German Reich. Hence, countless have to leave their country and run away to the United States, the United kingdom, Switzerland, or Sweden.

As mentioned earlier, it was already reckoned that if secondary neutrons were emitted through fission, the possible chain reaction would release enormous amounts of energy. Thus, bearing in mind the fact that German physicists were on the front line of uranium studies when Nazism spread throughout Europe, the fear that Germany would exploit an atomic bomb grew high among the community of scientific refugees. At the appeal of Hungarian physicist Leo Szilard, German theoretical physicist Albert Einstein, already in America, wrote a warning letter to President Franklin Roosevelt about the possibility of an atomic bomb as early as 1939. As an answer, modest financial support was decided for early uranium research at various universities, including studies on fission chain reactions, neutron emissions, and isotope separation [25].

In 1942, at the University of Chicago, researchers of the Chicago Met Lab (Metallurgical Laboratory) led by Enrico Fermi achieved the first self-sustaining chain reaction in the graphite pile known as CP-1. That experiment proved that nuclear energy could generate power, indicating an effective neutron multiplication factor of 1.0006[13] [26]. It also showed a viable method to produce plutonium. At this time, the United States had already entered the Second World War.

That was just the initial step in proving the bomb's feasibility. An exponentially growing number of neutrons, i.e., an exponential chain reaction going through enough reactions very quickly to release a lot of energy, was what was required for the possible bomb. The Manhattan Project, conceived next to research and design the atomic bomb, officially started at the end of the same year 1942. The American physicist J. Robert Oppenheimer was named director of a new laboratory to be built at Los Alamos, New Mexico, where numerous outstanding physicists, chemists, mathematicians, engineers, and explosives experts, among others, worked all together to develop such a weapon.

It is heartbreaking to recognize that such a great scientific breakthrough was associated with hundreds of thousands of lives lost because of the atomic bomb. This tragedy extends to the long-term radiation-related health effects experienced by the survivors and their descendants [27]. However, it is also undeniable that the advancements in nuclear sciences during this period have made substantial contributions to humanity's progress in the post-war era and beyond.

The efforts to construct the atomic bomb, which brought World War II to an end, marked a pivotal moment in scientific and technological advancement that distinguished the 20th century. What follows is a summary of the key events that led to remarkable developments in various fields such as human health, energy production, food, agriculture, and space exploration, all underpinned by the use of nuclear energy.

- The discovery of plutonium by American chemist G.T. Seaborg when bombarding uranium targets with deuterons[14] in a cyclotron accelerator [28, 29]. His findings not only opened the possibility for the synthesis of new artificial elements but also demonstrated that ^{239}Pu was more likely than ^{235}U to undergo fission, and could be produced from ^{238}U in a nuclear reactor.

- The construction of the world's first experimental nuclear reactor, Chicago Pile-1 (CP-1), which achieved a self-sustaining fission chain

---

[13] The neutron multiplication factor (k) is defined as the ratio of neutrons in generation (n+1) to the number of neutrons in generation n. If k = 1, then the chain reaction is self-sustaining. This state is called critical.

[14] Nucleus of deuterium $^{2}_{1}H$.

reaction for the first time. Designed by E. Fermi, it used natural uranium as fuel, graphite as a moderator, and cadmium rods to control the rate of neutrons [25].

- The development and enhancement of experimental techniques to measure fission cross-sections[15] with unprecedented accuracy over a wide neutron energy range. This also involved the compilation (collection, ordering, and transformation) of data files of evaluated nuclear reaction data [30]. Cross-sections are crucial to reactor design and control as they define parameters such as fuel requirements, fuel-moderator arrangements, and the number and size of control rods.

- The implementation of analytic formulas to compute the critical mass[16] of each ^{235}U and ^{239}Pu, with reduced uncertainties. Monte Carlo simulations were especially important for studying the probabilistic behavior of neutron transport in fissile materials and resolving critical dimensions, multiplication rates with or without tampers, etcetera [30, 31].

- The construction of two more experimental reactors: Chicago Pile-2 (CP-2) and Chicago Pile-3 (CP-3) based on the CP-1 model. CP-2 was operated at 1–10 kW and used for studies of neutron capture cross-sections, shielding, instrumentation, and as a training facility for production operations. CP-3 was the world's first heavy-water moderated reactor. It was used for research in reactor physics, fission product separation, tritium production experiments, and studies of radionuclide metabolism in laboratory animals [32].

- The experimentation and industrialization of three different methods of isotope separation to achieve the level of ^{235}U enrichment: an electromagnetic separation process, a gaseous diffusion process, and a thermal process. The electromagnetic and thermal processes were discontinued after the war ended.

- The construction of a second experimental air-cooled, graphite-moderated nuclear reactor to convert ^{238}U into ^{239}Pu using the neutrons emitted in the fission of ^{235}U. It was a prototype for the production-scale reactors that were built later (although with a different cooling system); was used to test the chemical separation facilities for plutonium in a pilot-scale arrangement; provided an

---

[15] Cross-section is the probability that a certain reaction, e.g., fission, will occur between two particles for a given set of parameters (energy), expressed in terms of area.

[16] Mass of fissile material to sustain the nuclear fission, which depends on its geometry, density, nuclear properties, enrichment level, whether there is a reflector or a tamper, and so on.

important experience for engineers, technicians, reactor operators, etc.; and supplied the Los Alamos laboratory with the first significant amounts of plutonium for the bomb design [25].

- The construction of three light water-cooled, graphite-moderated reactors, operating at 250 MW, for the production of ^{239}Pu. These reactors continued operating until 1989 and were responsible for the production of roughly two-thirds of the U.S. weapons stockpile of ^{239}Pu.

- The development of the first industrial-scale ^{239}Pu reprocessing method using a series of precipitation and solid-liquid phase separation. Extensive research on plutonium's chemistry preceded the design and construction of this process of several cycles of reduction, precipitation, phase separation, dissolution, and oxidation. Bismuth phosphate was used as a co-precipitate agent [33].

- Research into the properties and production of heavy water[17] (D_2O) aimed at its use as a reactor moderator, as well as the construction of four plants for its manufacture, three in the U.S. and one in Canada. The heavy water produced was used to build three reactors, the CP-3 already mentioned, and two experimental reactors at the Chalk River Laboratories near Ontario, Canada.

- The enhancement and development of measurement methods and instruments for radiation (scintillation, proportional and Geiger-Müller counters, ionization chambers, photographic, spectrographic, and activation methods, etc.), pointed toward process control and work protection, as well as to study the health effects and environmental consequences of radioactive materials.

References

[1] EPA U.S. Environmental Protection Agency. Technologically Enhanced Naturally Occurring Radioactive Materials (TENORM). U.S. Environmental Protection Agency, 12 July 2021. [Online]. Available: https://www.epa.gov/radiation/technologically-enhanced-naturally-occurring-radioactive-materials-tenorm. [Accessed 16 January 2022].

[2] International Atomic Energy Agency. 2001. IAEA Safeguards Glossary, International Nuclear Verification Series, IAEA/NVS/3, Vienna: IAEA, 2001.

[3] TWI Ltd. 2022. The Use of Thoriated Tungsten Electrodes. TWI Ltd, 2022. [Online]. Available: https://www.twi-global.com/technical-knowledge/faqs/faq-the-use-of-thoriated-tungsten-electrodes. [Accessed 2022].

[4] International Atomic Energy Agency. 2005. Thorium fuel cycle — Potential benefits and challenges, IAEA-TECDOC-1450, Vienna: IAEA, 2005.

---

[17] Heavy water is made with two deuterium ions (^2_1H) rather than two hydrogen ions and has a higher molecular weight than ordinary water.

[5] U.S. National Research Council Committee on the Biological Effects of Ionizing Radiations. 1988. Health Risks of Radon and Other Internally Deposited Alpha-Emitters, Washington: National Academies Press.

[6] International Atomic Energy Agency. 1996. Conditioning and Interim Storage of Spent Radium Sources IAEA-TECDOC-886, Vienna: IAEA.

[7] Ensslin, N. 2017. The Origin of Neutron Radiation. Los Alamos National Laboratory; Safeguards and Security Technology Training Program. [Online]. Available: https://www.lanl.gov/org/ddste/aldgs/sst-training/_assets/docs/PANDA/The%20Origin%20of%20Neutron%20Radiation%20Ch.%2011%20p.%20337-356.pdf. [Accessed 2 February 2022].

[8] Jing, K. 1999. Fission Probabilities, Fission Barriers, and Shell Effects Ph.D Dissertation, Berkeley: University of California.

[9] Urone, P. P. and R. Hinrichs. 2012. College Physics 31.7 Tunneling. OpenStax College, 21 June 2012. [Online]. Available: https://courses.lumenlearning.com/physics/chapter/31-7-tunneling/#:~:text=The%20probability%20of%20finding%20a,penetration%20or%20quantum%20mechanical%20tunneling. [Accessed 2 February 2022].

[10] U.S. NRC. 2021. U.S. Nuclear Regulatory Commission Glossary. U.S. Nuclear Regulatory Commission, 3 September 2021. [Online]. Available: https://www.nrc.gov/reading-rm/basic-ref/glossary.html#F. [Accessed 4 February 2022].

[11] Seaborg, G. T. 1963. Man-made Transuranium Elements. Englewood Cliffs: Prentice-Hall.

[12] U.S. National Research Council. 1996. Nuclear Wastes: Technologies for Separations and Transmutation, Washington: The National Academies Press.

[13] Shwageraus, E., P. Hejzlar and M. Kazimi. 2004. Use of thorium for transmutation of plutonium and minor actinides in PWRs. Nuclear Technology 147: 53–68.

[14] U.S. Department of Energy. 2021. Science Made Simple: What Are Nuclear Fusion Reactions? SciTechDaily, 26 May 2021. [Online]. Available: https://scitechdaily.com/science-made-simple-what-are-nuclear-fusion-reactions/. [Accessed 4 February 2022].

[15] International Thermonuclear Experimental Reactor (ITER). 2022. 60 YEARS OF PROGRESS. ITER Organization, 2022. [Online]. Available: https://www.iter.org/sci/BeyondITER. [Accessed 5 February 2022].

[16] Zylstra, A., O. Hurricane and D. Callahan. 2022. Burning plasma achieved in inertial fusion. Nature 601: 542–548.

[17] Gould, C. R., D. Davis, L. Wilets and P. J. Siemens. 2003. Nuclear Physics. pp. 721–738. *In*: Encyclopedia of Physical Science and Technology. Academic Press.

[18] Steere, A. R. 2005. A Timeline of Major Particle Accelerators (Thesis for Master Degree), East Lansing: Michigan State University, Department of Physics and Astronomy.

[19] Chadwick, J. 1932. Possible existence of a neutron. Nature 129: 312.

[20] Nesvizhevsky, V. and J. Villain. 2017. The discovery of the neutron and its consequences (1930–1940). Comptes Rendus Physique, no. Open access article under CC BY-NC-ND license.

[21] Science History Institute Chemistry- Engineering- Life Sciences. 2017. Irène Joliot-Curie and Frédéric Joliot. Science History Institute, 9 December 2017. [Online]. Available: https://www.sciencehistory.org/historical-profile/irene-joliot-curie-and-frederic-joliot. [Accessed 10 February 2022].

[22] APS Physics. 2007. This Month in Physics History December 1938: Discovery of Nuclear Fission. American Physical Society, December 2007. [Online]. Available: https://www.aps.org/publications/apsnews/200712/physicshistory.cfm. [Accessed 11 February 2022].

[23] Von Halban, H., F. Joliot and L. Kowarski. 1939. Number of neutrons liberated in the nuclear fission of uranium. Nature 143: 680.

[24] Bohr, N. and J. A. Wheeler. 1939. The mechanism of nuclear fission. Physical Review 56: 426.

[25] U.S. Department of Energy. 2022. The Manhattan Project an interactive history. U.S. Department of Energy - Office of History and Heritage Resources [Online]. Available: https://www.osti.gov/opennet/manhattan-project-history/Events/1942-1945/rivals. htm. [Accessed 14 February 2022].

[26] Atomic Heritage Foundation. 2016. "Chicago Pile-1," Atomic Heritage Foundation in partnership with National Museum of Nuclear Science & History, 1 December 2016. [Online]. Available: https://www.atomicheritage.org/history/chicago-pile-1. [Accessed 17 February 2022].

[27] Samet, J. M. and O. Niwa. 2020. At the 75th anniversary of the bombings of Hiroshima and Nagasaki, the Radiation Effects Research Foundation continues studies of the atomic bomb survivors and their children. Carcinogenesis 41(11): 1471–1472.

[28] Hoffman, D. C. 1996. The Transuranium Elements: From Neptunium and Plutonium to Element 112. In Conference Proceedings NATO Advanced Study Institute on "Actinides and the Environment", Chania, Crete.

[29] Seaborg, G., E. M. McMillan, J. W. Kennedy and A. C. Wahl. 1946. Radioactive Element 94 from Deuterons on Uranium. Physical Review 69: 366–367.

[30] Chadwick, M. B. 2021. Nuclear Science for the Manhattan Project and Comparison to Today's ENDF Data. Nuclear Technology 207(sup 1): S24–S61.

[31] Sood, A., R. A. Forster, B. Archer and R. Little. 2021. Neutronics calculation advances at Los Alamos: Manhattan project to Monte Carlo. Nuclear Technology 207(sup 1). https://doi.org/10.1080/00295450.2021.1956255.

[32] Nuclear Engineering Division Argonne National Laboratory. 2022. Reactors Designed by Argonne National Laboratory. Early Exploration. Argonne National Laboratory, [Online]. Available: https://www.ne.anl.gov/About/reactors/early-reactors.shtml. [Accessed 16 February 2022].

[33] Schwantes, J. and L. Sweet. 2011. Contaminants of the Bismuth Phosphate Process as Signifiers of Nuclear Reprocessing History PNNL-21057. Pacific Northwest National Laboratory, Richland.

CHAPTER 2

Atoms for Peace

The scientists involved in the Manhattan Project, even amidst the turmoil of war, believed that fission energy had greater applications. They saw its potential in generating heat and electricity, and also in creating radioactive elements for use in medicine, industry, and research [1].

Almost all of the scientific community after WWII agreed that the vast resources used for weapons had to be repurposed for peaceful applications. The first ideas were to harness the heat produced in the process either for direct use—e.g., nuclear marine propulsion—or for generating electricity. Marine propulsion plants offered long intervals of operation before refueling with no cargo or supplies space for the fuel on the vessel. Accordingly, the design, development, and production of nuclear marine propulsion plants started early in the 1940s in the US. However, their use was limited to naval warships, such as nuclear submarines, aircraft carriers, destroyers, and cruisers [2]. In civil merchant ships, nuclear marine propulsion had not succeeded beyond a few experimental ships like the NS Savannah built in the US, and launched in 1962, and the Russian icebreakers Lenin, launched in 1959, and NS Arktika, launched in 1975, which also was the first surface vessel to reach the North Pole in 1977 [3].

First post-war nuclear reactors and electricity production

The entry into operation of the first non-military reactors is shown in the graphic timeline in Figure 2.1. They are now known as first-generation reactors.

The small experimental breeder reactor (EBR-I), designed and operated by Argonne National Laboratory in Idaho, was the first nuclear reactor to produce electricity. At start-up in December 1951, it powered four 200-watt lightbulbs as illustrated in Figure 2.2, and eventually generated enough electricity to light the entire facility [4]. This pioneering

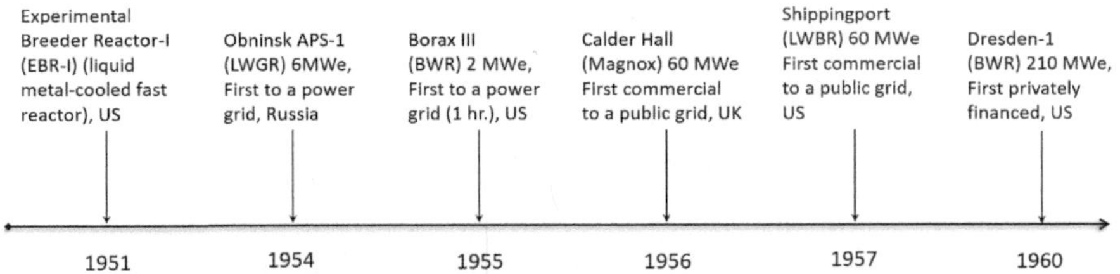

First Generation Reactors

Figure 2.1. First-generation non-military nuclear reactors' start-up timeline.

Figure 2.2. Experimental Breeder Reactor (EBR-1), a US National Historic Landmark near Arco, Idaho (photo courtesy of US Department of Energy).

fast reactor used liquid metal for cooling and was built to confirm that it could generate more fuel than it consumed.

In Russia, the graphite-moderated reactor to produce plutonium at the Institute of Physics and Power Engineering in Obninsk was modified for electricity and heat generation. As a result, in June 1954, the first nuclear-powered electricity generator started its operation connected to a power grid. This reactor, illustrated in Figure 2.3, was a Light Water Graphite Reactor (LWGR) with a design capacity of 6 MWe. It operated for the Mosenergo grid for three years. Afterward, it was used for research and isotope production [5].

Figure 2.3. First nuclear plant to connect to the grid in the world, Obninsk (photo courtesy of US Department of Energy).

The BORAX[1] III, an experimental boiling water reactor (BWR) with a capacity of 2 MWe, was the first nuclear reactor to connect to the grid in the US, in 1955. It provided the town of Arco, Idaho, with complete electrical service for over an hour [6]. The so-called BORAX experiments have carried significant importance in the later development and marketing of full-sized commercial BWR power plants that were based on the technical information revealed by these experiments [7].

The Calder Hall reactor illustrated in Figure 2.4, located at Sellafield, Cumbria in the UK, was the first nuclear power plant in the world used for commercial electricity generation. Calder Hall was designed to produce plutonium for military applications, as well as electricity. The plant consisted of a Magnox[2] graphite-moderated, CO_2-cooled reactor, using natural uranium fuel elements covered by a Magnox sheath. With a gross capacity of 60 MWe, Calder Hall was first connected to the grid in 1956 and operated for 47 years [8]. It holds the record for being the oldest power plant [9]. In comparison to the present day, where nuclear power contributes only 16% of the UK's electricity generation, back in 1968, nuclear power stations were responsible for approximately 25% of the country's total annual electricity production [10].

The Shippingport Atomic Power Station in the US, illustrated in Figure 2.5, was the first full-scale nuclear power plant in the world devoted exclusively to peacetime uses. Its first commercial power was

---

[1] Boiling Reactor Experiments (BORAX).

[2] The Magnox name refers to the magnesium alloy wrapping each uranium fuel rod.

Figure 2.4. Calder Hall at Sellafield soon after opening in the 1950s (photo courtesy of UK government agencies, licensed under OGL v.3).

Figure 2.5. Shippingport, the first full-scale nuclear power generating station in the US (photo courtesy of the US Department of Energy).

produced and fed into the grid for the Pittsburgh area in 1957 [11]. This water breeder reactor of 60 MWe, with a seed-and-blanket core type, used highly enriched uranium (93% ^{235}U) as seed fuel, surrounded by a blanket of natural ^{238}U [12]. The reactor was designed by the Westinghouse Electric Corporation in cooperation with the Division of Naval Reactors of the Atomic Energy Commission and was operating for 25 years [13].

Dresden-1, designed by General Electric, was the first privately financed nuclear power plant built in the US. It started operation in 1960 under Exelon Generation Co., LLC. Dresden-1 boiling water reactor (BWR), with a capacity of 250 MWe, generated electrical power for 18 years and was permanently shut down in 1978 [14].

Creation of the IAEA to promote the peaceful use of nuclear energy

The US was not the only country to be involved with nuclear weapons in the 40s and early 50s. Wary of the US plans to build an atomic bomb, as early as 1942, Russia approved a research program to achieve a controlled chain reaction. Under the leadership of nuclear physicist Igor Kurchatov, the Russian first reactor came operational in 1948, and the next year, in 1949, it performed its first nuclear test in Semipalatinsk, Kazakhstan [15].

The UK followed and was the third country to test nuclear weapons. Although British scientists had played a crucial role in the success of the Manhattan Project, the UK decided to carry out its atomic program independently because of the secrecy of nuclear weapons information imposed by the American Atomic Energy Act of 1946. The first UK reactor went critical in 1950 and the reprocessing plant to provide fissile material started operating in 1951. The first British nuclear test was conducted at the Montebello Islands in Western Australia in 1952 [16].

The race for nuclear weapons did not stop there. On the contrary, just as soon as Britain had successfully tested its fission device, the US and Russia produced their first hydrogen bombs. In 1952, the first hydrogen bomb was detonated on the Eniwetok Atoll in the Pacific Marshall Islands. In 1953, Russia announced it had tested its first hydrogen bomb in central Siberia [17].

The first French reactor was built at Marcoule, in the south of France, and went critical in 1956. Afterward, the first French nuclear test took place in the Sahara Desert, near the town of Reggane, Algeria in 1960. It is worth mentioning that Frédéric Joliot-Curie, the son-in-law of Marie Curie, was a pivotal personality in restarting atomic research in France after WWII. In 1945, he was named the high commissioner in charge of all scientific and technical work at the Commissariat of Atomic Energy (CEA), although he was dismissed in 1950 given his opinion against all nuclear weapons [18].

Having access to uranium resources, China went on with its plans to build an atomic bomb without a nuclear plant. The Institute of Physics and Atomic Energy in Beijing was responsible for the research on bomb design [19]. The Chinese also constructed a uranium enrichment plant to produce the weapon-grade uranium they needed [20]. The bomb materials were then assembled at the Lop Nur Nuclear Weapons Test Base. The first Chinese test was conducted at the Lop Nur test site in the Gobi Desert of Xinjiang province, Western China, in 1964 [21].

The secrecy that once surrounded the development and possession of nuclear weapons had been lifted already. In response to growing fears about their proliferation, the US, the UK, and Canada decided to negotiate with Russia. Their goal was to establish an international agreement that would control the increase of nuclear weapons and promote technical cooperation through an intergovernmental organization known as the United Nations (UN). This organization was created in 1945 to maintain global peace. A significant contribution to this agreement was President Eisenhower's address to the UN General Assembly in December 1953, known as Atoms for Peace [22].

To halt the proliferation of nuclear weapons and promote a safe, secure, and peaceful balanced application of nuclear energy over the world, in 1957, the United Nations created the International Atomic Energy Agency (IAEA) as an organization independent from any government [23]. The IAEA was established with a dual mission. On one hand, it should aim at preventing nuclear energy from being used for military purposes; and, on the other, ensure that nuclear energy for peaceful purposes is used in the safest possible way [24].

Currently, 173 countries and the Holy See (Vatican City) are Member States of the IAEA. Furthermore, thanks to technical cooperation and non-proliferation efforts, there are 442 nuclear power reactors in operation in 32 countries, with a total capacity of 392.6 GWe [25], and 19 countries, have 52 reactors under construction with a total capacity of 54.4 GWe. Among these are Bangladesh and Turkey who are building reactors for the first time.

Present worldwide efforts in the peaceful use of nuclear energy are reflected in the data from the IAEA 2020 [26] annual report presented in Table 2.1. As seen, the number of countries and territories receiving support through technical cooperation programs reached 146, including 35 least-developed countries (e.g., Angola, Bangladesh, and Senegal), where 1329 facilities are under safeguards agreements with the IAEA for verification of their compliance towards preventing any military purpose. 225,000 significant quantities[3] of nuclear materials were accounted for and backed by containment and surveillance measures in those countries [27]. In addition, the IAEA obtained from the Member States qualified experts and resources, e.g., to support 1400 regional or multinational training courses around the world.

---

[3] One significant quantity is the approximate amount of nuclear material for which the possibility of manufacturing a nuclear explosive device cannot be excluded.

Table 2.1. IAEA activities overview.

Countries, territories assisted	International laboratories	Multilateral conventions	Facilities under safeguards	Significant quantities	Coordinated research projects	Training & education courses
146	15	11	1329	225,000	137	1400

Technical Cooperation Projects (TCP) are the means the IAEA uses to assist developing countries with resources and experts to address national or regional key priorities, including personnel training courses and education. They can assist, for example, in the development of the safety and security regulatory framework, in building competence in safety and security, healthcare applications, nutrition improvements, and so on. TCPs can be approved for one country or region. For instance, AFRA (the African Co-operative Agreement for Research, and Development) and ARCAL (for the Promotion of Nuclear Science and Technology in Latin America and the Caribbean) are regional projects. The areas of cooperation include energy, food and agriculture, health and nutrition, safety and security, water and the environment, industrial applications and radiation technology, as well as nuclear knowledge development and management [28].

Through Coordinated Research Projects (CRP), the IAEA brings together scientists and resources from several developed countries to collaborate on a focused research topic of common interest. They are usually approved for 3–5 years and involve 8–15 laboratories, research teams, or institutions that do not receive supporting funding. For example, a project was approved in 2022 to coordinate the research activities on the impurities intentionally injected into the plasmas of experimental nuclear fusion devices.

Other IAEA activities include multilateral conventions, i.e., agreements between member states, less formal than treaties, to cover particular matters [29]. For example, a collaborative team of experts from 62 countries adopted two conventions after the Chornobyl accident in 1986. They were the Convention on Early Notification of a Nuclear Accident which establishes the obligation of prompt notification concerning any release of radioactive material that could have a transboundary impact [30]; and the Convention on Assistance in the Case of a Nuclear Accident or Radiological Emergency which facilitates quick assistance in case of a nuclear accident or a radiological emergency to minimize its consequences and to protect life, property, and the environment [31].

Nuclear material definition: source material and special nuclear material

The definition of nuclear material is ruled by the Non-Proliferation Treaty (NPT); the IAEA safeguards system; and the corresponding laws and regulations of Member States parties to the Treaty. The definition primarily includes two elements: source material and special nuclear fissionable material.

According to Article XX of the IAEA Statute [32], the source material is uranium containing the mixture of isotopes occurring in nature; uranium depleted in the isotope 235; thorium; and any of them in the form of metal, alloy, chemical compound, or concentrate. This concept does not apply to ore or ore residue. Special nuclear fissionable material refers to ^{239}Pu (plutonium-239); ^{233}U (uranium-233); uranium enriched in ^{235}U or ^{233}U; or any material containing one or more of the before-mentioned radionuclides. It does not include the source material.

The definition of nuclear material in the US Atomic Energy Act of 1954 [33] included byproduct materials along with source material and special nuclear material. Source material, according to the Nuclear Regulatory Commission (NRC) [34], is thorium or uranium with a ^{235}U content equal to or less than that found in nature ($\leq$ 0.71% ^{235}U); and special nuclear material is plutonium, ^{233}U, or uranium with a ^{233}U or ^{235}U content greater than that found in nature (> 0.71% ^{235}U). Byproduct material is currently defined under the title Regulated Materials and not under nuclear materials.

The Treaty Establishing the European Atomic Energy Community (Euratom) classifies nuclear materials as ores, source materials, or special fissile materials [35]. Hence, special fissile materials mean ^{239}Pu; ^{233}U; and uranium enriched in ^{235}U or ^{233}U, and do not include source material. Uranium enriched in ^{235}U or ^{233}U means uranium containing ^{235}U or ^{233}U, or both, in such an amount that the abundance ratio of the sum of ^{235}U and ^{233}U to ^{238}U is greater than the ratio of ^{235}U/^{238}U in nature. Source material means uranium containing the mixture of isotopes occurring in nature; uranium whose content in ^{235}U is less than that in nature; thorium; or any of them in the form of metal, alloy, chemical compound, or concentrate. Ores mean any ore that contains substances from which the source material defined above may be obtained by the appropriate chemical and physical processing.

The Non-Proliferation Treaty, safeguards, and inspection

The Non-Proliferation Treaty (NPT) represents a historic breakthrough in the global effort to take advantage of the benefits of nuclear energy. It is an agreement between the five acknowledged Nuclear Weapons States—

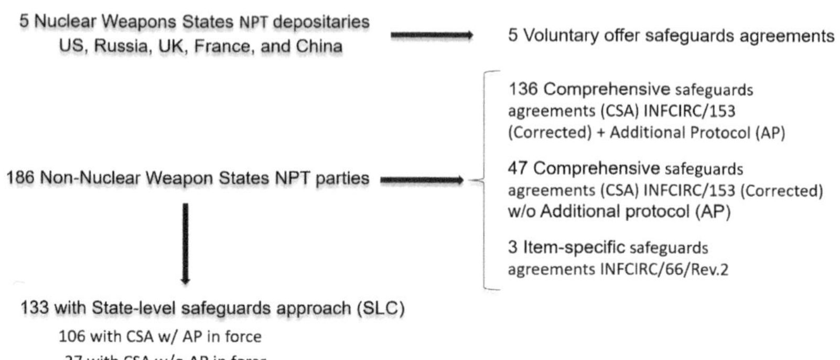

Figure 2.6. Safeguards agreements between non-nuclear weapons states and the IAEA.

US, Russia, UK, France, and China that had manufactured and tested a nuclear weapon before 1967—and other non-nuclear-weapon states interested in nuclear technology. The United Nations General Assembly adopted the NPT in 1968, and the Treaty entered into force in 1970 after 43 States including the five depositaries ratified it. In 1995, the NPT was extended indefinitely. As of 2022, 191 States were parties to the NPT. A summary is shown in Figure 2.6.

The central idea of the NPT is that assistance and cooperation in nuclear energy and technology must be exchanged for guarantees, under IAEA scrutiny, that no facility or material would be diverted to develop and create nuclear weapons [36]. This implies two main responsibilities for the State. First, it must place all source or special fissionable materials in all peaceful nuclear activities under safeguards—whether these activities occur within its territory, under its authority, or under its control anywhere. Second, it must accept taking on reporting and record-keeping responsibilities before the IAEA and several verification actions, including allowing the IAEA to review the design information and conduct inspections.

Because of the previous commitments, each State party of the NPT is required to enter into safeguards agreements with the IAEA to explicitly declare its guarantees and authorize the IAEA to verify the accomplishment of its responsibilities. These agreements are called comprehensive safeguards agreements (CSAs). The numbers in Figure 2.6 show that, at the end of 2020, 183 States had CSAs with the IAEA under the structure and content of INFCIRC/153 [37], and three of them: India, Israel, and Pakistan, which are non-NPT parties, had item-specific safeguards agreements as per INFCIRC/66/Rev.2 [38].

Nuclear weapons States are not required to have safeguards agreements with the IAEA. However, from the early 80s, the five of them have concluded voluntary offer agreements (VOA) and additional protocols to facilitate gaining experience. Under these VOA, nuclear weapon states, as depositaries of the NPT, are free to choose which facilities would be under safeguards and which withdrawn from safeguards. They can also add or remove nuclear material [27].

Additional Protocols (AP), as per INFCIRC/540 [39], are instruments to grant the IAEA complementary inspection authority to confirm both declared and potential undeclared activities. They provide additional tools to verify the accuracy and completeness of the declared information, confirm the absence of undeclared nuclear material, and resolve any questions or inconsistencies. Some of those measures are, for example, early provision of design information; enhanced inspector access from mines to nuclear waste; reports including exports and imports of nuclear material; two-hour short-notice inspector access to all buildings on a nuclear site; reports on the manufacturing and export of non-nuclear, but nuclear-related, equipment and material; locations of nuclear fuel cycle-related research and development activities not involving nuclear material; environmental sampling at the site; and so forth.

It is worth mentioning that the scope of non-NPT agreements of type INFCIRC/66/Rev.2 has been expanded to include provisions for safeguarding non-nuclear materials related to nuclear facilities, such as heavy water and zircaloy, as well as non-nuclear facilities like heavy water production plants [40].

The State Level Concept (SLC) implemented for 133 countries as of 2020, is an advanced approach to safeguards that requires particular objectives at the country and facility level. The verification is then conducted from the top down, and the conclusions are the results of the annual verification plan for each country based on such objectives, some specific factors of each country, and all the safeguards-relevant information[4] available to the IAEA [41, 42].

To implement the state-level approach, the IAEA has set up a total of 100 multidisciplinary State Evaluation Groups (SEGs) with appropriate expertise. Apart from identifying the country's objectives, updating its evaluation with new information on an ongoing basis, preparing an annual evaluation, and recommending conclusions for the country's safeguards to the IAEA senior management, these groups, through tools

---

[4] Safeguards-relevant information includes the information from the State itself; the information from the safeguards activities conducted by the IAEA in the field and at Headquarters; and other relevant information obtained from open sources and third parties.

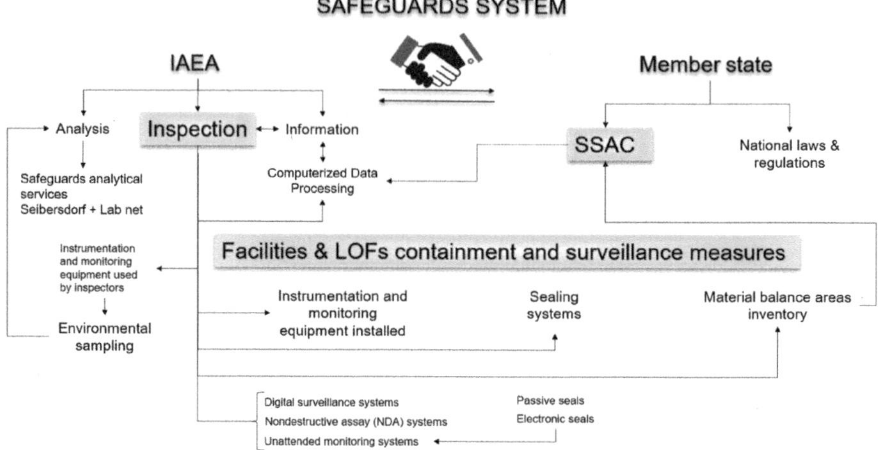

Figure 2.7. The Non-Proliferation treaty's verification mechanism.

like Acquisition Path Analysis,[5] can identify all technically plausible paths by which a particular country, with its technical capabilities, could acquire nuclear material suitable for a nuclear weapon [43, 44].

The main components of the Safeguards System, the verification mechanism of the NPT, are illustrated in Figure 2.7. This mechanism is a cooperative agreement between the IAEA and its Member States, where both parties should follow the same strategy and plan of action to implement the corresponding safeguards obligations, without hampering or interfering with the economic and technological development of the country.

As shown in Figure 2.7, the NPT verification mechanism relies on three basic elements. The first is the accounting system, known as the System of Accounting for and Control of Nuclear Material (SSAC), which is established by law at both the state and facility levels to locate and quantify inventories and flows of nuclear materials that are required to be declared. The second element involves IAEA inspection tasks that verify the correctness and completeness of this information, including sample collection by inspectors and measurement at laboratories. The third element is the use of fixed surveillance and containment systems provided by the IAEA to prevent and monitor any diversion of nuclear materials.

---

[5] It is a structured method to analyze the technically plausible paths for nuclear weapon acquisition.

Control and accountability of nuclear material

Each country that is a party to the NPT is required to provide information on facilities and locations outside of facilities (LOFs) where nuclear materials are stored, processed, or used. Similarly, these countries must declare nuclear materials, nuclear material flows, inventories, and inventory changes in each of these facilities or LOFs. Below are some figures of the facilities and LOFs that were under safeguards worldwide in 2020 [28]:

Facilities	Amount
Power reactors	263
Research reactors and critical assemblies	150
Conversion plants	17
Fuel fabrication plants	41
Reprocessing plants	11
Enrichment plants	19
Separate storage facilities	139
Other facilities	77
Locations outside facilities[6]	604
Total facilities and locations outside facilities	1321

To implement the mechanisms for accounting, monitoring, reporting, and granting access to nuclear material to IAEA inspectors, the country must first enact the corresponding laws and regulations to facilitate the IAEA verification activities and to establish the national regulatory and supervisory authorities.

Under the provisions of the safeguards agreements with the IAEA [45], the Member State should also provide the starting point for the application of the IAEA safeguards. This means that the nuclear materials to be under safeguards are initially declared. In addition, the Member State should provide the IAEA with design information on facilities and LOFs where nuclear materials will be stored, processed, or used. At planned intervals, the Member State should supply the IAEA with information on nuclear material inventories and inventory changes concerning each declared facility or LOF.

As seen in Figure 2.8, all these involve a legal framework (laws and regulations), a recognized authority — often the same established to enforce compliance with national and international requirements for the safety

---

[6] Includes 65 locations in states with amended small quantities protocols, i.e., minimal nuclear material.

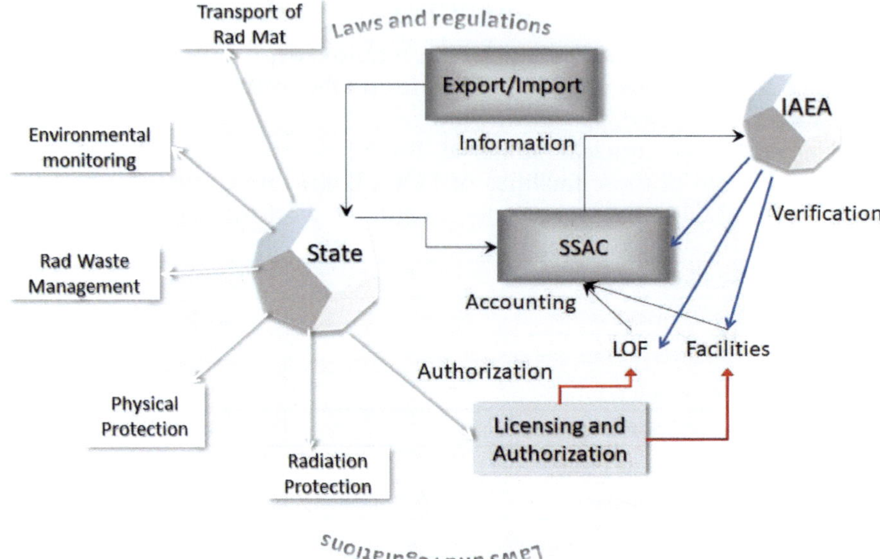

Figure 2.8. Connections between different parts of the State System of Accounting for and Control (SSAC) for nuclear materials and the IAEA.

of radiation sources—and a management system. The last may include several national bodies and institutions associated with the transport and import/export of radioactive materials, environmental radiation monitoring, radioactive waste management, physical protection, and licensing of facilities and locations where nuclear materials are produced, used, and stored [46].

For example, in the US, the Nuclear Regulatory Commission's (NRC) Office of Nuclear Material Safety and Safeguards (NMSS) is responsible for the licensing and regulation of facilities and materials associated with the use, processing, transport, and handling of nuclear materials, including uranium recovery activities and the fuel in commercial nuclear reactors [47]. The Code of Federal Regulations (CFR) Title 10 Part 75 regulates the implementation of the safeguards agreements between the US and the IAEA [48], and the CFR Title 10 Part 74 regulates the material control and accounting of special material [49].

The mentioned measures should ensure three important tasks that are shown on the right side of Figure 2.8. These are, the flow of information on nuclear materials from the facilities and/or LOFs, including export and imports; the authorization of the facilities and nuclear materials by the national authority; and the process of verification by the IAEA.

Authorization means that only specific facilities or locations are permitted, or licensed, by the Member State to hold nuclear material and only for definite purposes. These last could be possession, use, acquisition, import/export, disposal, isotope separation, or enrichment, subject to the conditions specified by the corresponding laws and regulations.

IAEA verification activities such as account and record auditing, inspections, and visits directly to facilities and LOFs are also indicated in Figure 2.8. The flow of nuclear material accounting information between the State, the facilities and LOFs, and the IAEA, including import/export information is shown through the SSAC.

The national authority is independent of any operators or promoters of nuclear energy. It controls the licenses or permits; oversees the nuclear material accounting activities at the facility and LOF levels; runs the SSAC at the national level; submits information on nuclear materials and facilities to the IAEA, and enables the IAEA inspectors the proper access to the facilities and LOFs.

To declare the quantities of nuclear material received, produced, shipped, lost, or otherwise removed from inventory, and the quantities on the inventory—including exports and imports, the SSAC should work on a structure of material balance areas.[7] Therefore, nuclear materials accounts, records, and reports are created at agreed material balance areas in facilities and LOFs, including the receipts into and the transfers out. It is common, except for complex facilities, that a facility and a material balance area coincide. Also, it is possible to combine various LOFs in one material balance area.

Exports and imports include non-nuclear material and equipment related to the nuclear fuel cycle, such as deuterium and heavy water, nuclear-grade graphite, reactor control rods, zirconium tubes, and the like.

The SSAC at the national level also checks the measurement accuracy, precision, and uncertainties for the reported quantities. To do so, it promotes creating or using, the required metrological organizations to ensure the use of the most current international standards for measurement systems, and the calibration of the corresponding equipment at appropriate intervals.

The procedures to prepare and submit detailed and timely reports to the IAEA are also responsibilities of the SSAC. Amid procedures are those aimed at fulfilling important goals of the accountability of nuclear materials. That is, taking a physical inventory;[8] identifying, reviewing, and evaluating differences in the shipper/receiver measurements,

---

[7] Material balance area means an area in or outside of a facility where: (i) the quantity of nuclear material in each transfer into or out can be found out; and (ii) the physical inventory of nuclear material can be taken when necessary, following specified procedures.

[8] Physical inventory means the sum of all the measured or derived estimates of batch quantities of nuclear material on hand at a given time within a material balance area, obtained by specified procedures.

including receipts into and transfers out; and evaluating accumulations of unmeasured inventory and unmeasured losses.

Examples of reports are the inventory change report, the material balance report, and the physical inventory listing. The material balance report is submitted at the end of the material balance period for each material balance area. It includes the shipper receiver difference, if applicable, and the material unaccounted for, if any. The material unaccounted for is kept to the lowest practicable level, i.e., small enough to be within the uncertainty of the eventual material balance established for each facility.

Depending on the facility type, e.g., nuclear reactor, fuel fabrication, or enrichment facility, the nuclear material may be in bulk form—e.g., powder, liquid, or fuel pellets—or contained in several separate items, e.g., fuel assemblies. The basic unit of nuclear material accounting in reports sent to the IAEA is a batch, i.e., a portion of nuclear material handled as a unit for accounting purposes, which is located in the same key measurement point[9] and has the same physical and chemical characteristics defined by specifications or measurements [45].

To quantify the nuclear material received, produced, shipped, lost, or otherwise removed from inventory, there should be key measurement points, equipment, and sampling and measurement methods, both chemical and non-destructive. A very simplified scheme of key measurement points in the material balance area of a nuclear reactor is illustrated in Figure 2.9. Key measurement points of this simple scheme are at the receipt of fresh nuclear fuel (where the material is stored at arrival), at the reactor core (where the material is processed), and at the spent nuclear fuel storage (where the material is finally transferred). Some measurement methods are also given in this example.

Databases using relevant computer hardware and software have been developed to manage all the above-mentioned data. The extent of the requirements and the different levels of computerized information systems for recording and reporting depends on the type of nuclear activity and the form and quantity of nuclear material [50]. For example, a country with minimal nuclear material probably only might need a spreadsheet-type system with manual reporting to notify occasional import/export operations and the annual inventory changes. Conversely, a country with several facilities, including at least one commercial power reactor under safeguards, might need commercially available information

---

[9] A key measurement point is a location where there is nuclear material, in such a form, that it may be measured to determine material flow or inventory, e.g., the inputs, outputs, and storage areas. The IAEA and the State agree on the flow and inventory key measurement points at a facility.

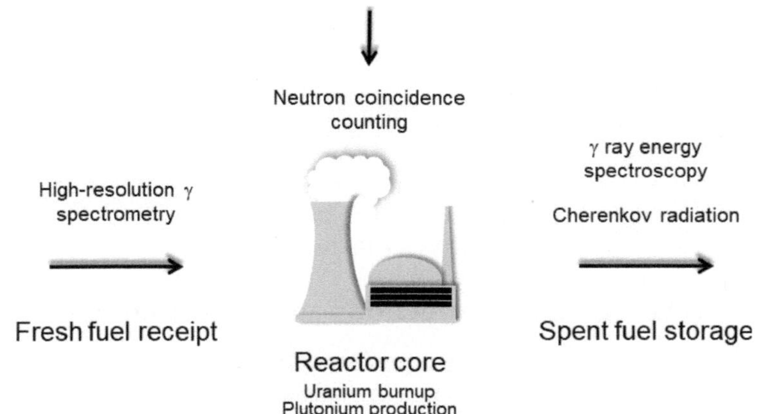

Figure 2.9. A simplified example of key measurement points in a nuclear reactor.

Table 2.2. Significative quantities and timeliness (Data from [27, 52]).

Special Nuclear Material	Significant Quantity (SQ)	Timeliness
Plutonium (< 80% ^{238}Pu)	8 kg total Pu	Irradiated: 3 months Nonirradiated: 1 month
Highly Enriched Uranium (≥ 20% ^{235}U)	25 kg ^{235}U	Irradiated: 3 months Nonirradiated: 1 month
Low Enriched Uranium (< 20% ^{235}U)	75 kg ^{235}U	12 months
Natural uranium	10 t	12 months
Depleted uranium	20 t	12 months

systems for nuclear material accounting at the national and facility levels, like the NAC Reporter™ from NAC International [51].

As a general rule, the goal of material accounting is to be able to detect the diversion of a minimum of one significant quantity (SQ) within the timeliness interval corresponding to the given category of nuclear material shown in Table 2.2. The timeliness goal reflects the time required to convert material diverted from a given facility into a weapons-usable quality [52]. The IAEA may establish different quantitative criteria for timeliness and significance based on the effectiveness of each national SSAC.

Process monitoring with data transmission and increased digitalization of nuclear material accounting are technologies already deployed by the IAEA to counteract the possibility of undeclared enrichment or reprocessing activity where fissile materials or separated plutonium can be produced, or where hold inventories of direct-use materials could be diverted very quickly. Other technologies under consideration include wide-area environmental monitoring [53], antineutrino detection [54], and satellite-based multispectral analysis [55].

Widespread computer usage has also been adopted by the IAEA over the years. There are automated operations and measurements for controlling nuclear material and for facilitating the flow of information reported by states to the IAEA, including the introduction of computerized inspection reports. Several ways were also developed for inspectors to receive and process nuclear material accountancy information from operator computers directly into IAEA portable computers with greater accuracy and requiring less effort by facility operators and inspectors.

Safeguards inspections and surveillance

As required by safeguards agreements, the IAEA has both the right and obligation to verify the information reported and the declarations made by Member States. Verification measures include *on-site* inspections, visits, ongoing monitoring (both remote and *in situ*), and evaluation by the SEG.

There are three types of inspections: *ad-hoc*, routine, and special. *Ad-hoc* inspections are performed to check on the initial report of nuclear material; on reports on changes to the initial report; and also, to verify the nuclear material involved in international transfers. Routine inspections are the most frequently used and are limited to strategic points.[10] They are performed according to a schedule or might be of unannounced or short-notice character. Special inspections are performed to verify the information provided in special reports, or they can be carried out if the IAEA has reason to believe that the State was performing unreported nuclear activities, or that the information received from the State and obtained in routine inspections does not match. In both cases, inspector access to locations that are not declared by the State is permissible, subject to consultations between the IAEA and the State [56].

IAEA's special inspection rights include not only access to declared nuclear material, but also to locations where there are indications of the presence of nuclear material and, what is more, to information and locations where there is no indication of the presence of nuclear material if necessary, e.g., research and development activities related the uranium enrichment and fuel reprocessing where no nuclear material is present [57].

Safeguards visits are also made to declared facilities and LOFs at appropriate times during the lifecycle to verify the safeguards-relevant design information. For instance, during construction, to establish the completeness of the declared design information; during routine operations and following maintenance, to confirm that no modification

---

[10] Strategic point is a location within a nuclear facility, or other location containing nuclear material, through which nuclear material is expected to flow.

was made; and during a facility decommissioning, to confirm that sensitive equipment was rendered unusable.

During, and in connection with, on-site inspections or visits at facilities and LOFs, inspectors might perform auditing of accounting and operating records and compare these records with the State's accounting reports to the agency; verify the nuclear material inventory and inventory changes; perform a visual examination, nondestructive analysis, weighing, or volume measurements, and take samples of nuclear materials for laboratory analysis; collect environmental samples to identify signs of the existence of undeclared nuclear material and activity; and apply containment and surveillance measures (e.g., seals, mount cameras, etc.) to detect unreported movement or tampering with nuclear materials. Figure 2.10 shows an inspector installing a surveillance camera.

To properly detect any diversion or misuse of declared nuclear material, declared facilities and LOFs are supplied with nondestructive analysis, containment, and surveillance systems.

As the examples in Table 2.3 show, the instruments used for non-destructive analysis include radiation detectors to measure neutron and gamma radiation, and various sensors to monitor temperature, flow, and other parameters. Analysis ranges from quantitative verification of enrichment levels to the purely qualitative detection of plutonium and uranium in fresh and spent fuel, and the presence of nuclear material in general [58]. Detectors can be part of small portable units, like those used

Figure 2.10. Inspectors mounting a surveillance unit. Slovakia's Mochovce NPP (photo courtesy of IAEA under CC BY-SA 2.0).

Table 2.3. Some methods used for nondestructive analysis in safeguards verification.

Nondestructive analysis method	Use example
Low and medium resolution γ spectrometry	Verify U enrichment, spent fuel, and Pu isotopic composition
High-resolution γ spectrometry	Determine the ^{235}U enrichment of uranium hexafluoride (UF$_6$) in shipping cylinders
Gross neutron counting	Check the presence of fissile nuclear material in non-irradiated fuel
Neutron coincidence counting	Determine Pu and ^{235}U content
Multiplicity coincidence counting	Determine ^{240}Pu effective mass
Gross neutron and gamma-ray detection	Spent fuel from boiling water reactor (BWR) and pressurized water reactor (PWR)
Gamma-ray energy spectral analysis	Fission product γ rays in spent fuel assemblies — either from ^{137}Cs (> four years cooling) or from ^{95}Zr/^{95}Nb (shorter cooling times)
Gamma-ray intensity scanning	Verify γ rays in storage baskets or stacks with irradiated CANDU fuel bundles — either from ^{137}Cs (> two years cooling) or from ^{95}Zr/^{95}Nb (shorter cooling times)
Neutron coincidence	Characterize spent fuel from research reactors stored underwater
Cherenkov radiation	Quantify the Cherenkov glow from spent fuel assemblies as a function of irradiation history and cooling time.
X-ray fluorescence analysis	Measure the enrichment of UF$_6$ gas in a pipe
Physical property measurement	Verify, e.g., the weight of an object, the wall thickness of a container, physical sizes, electromagnetic reflections of structural elements, and the liquid level in a tank

by safeguards inspectors during verification activities, or a component of installed systems for continuous unattended monitoring, which can or cannot be interfaced with a communication system.

Containment is the structural feature of a facility, container, or equipment that is used to corroborate its physical integrity. Likewise, to maintain the continuity of knowledge[11] by preventing undetected access to, movement of, or interference with the declared nuclear materials. The collection of information through an inspector and/or instrumental observation aimed at detecting movements, or tampering with equipment, samples, and data is called surveillance [59].

---

[11] Continuity of knowledge (CoK) is the outcome of a system of data or information regarding an item or activity that is uninterrupted and authentic and ensures the IAEA no changes have occurred between inspections.

Containment and surveillance techniques can be single or dual and are typically set up to cover any plausible diversion path. They can be, for example, a seal with or without remote monitoring capability, an unattended radiation detector, or an optical surveillance camera such as the digital camera module DCM-C5, an image-generating device for motion detection with several digital inputs for external triggers. Dual systems are used for difficult-to-access locations and consist of, at least, two functionally independent devices, with tampering and failure modes based on different physical principles.

Enclosures containing nuclear materials or protecting signals from unattended monitoring systems and surveillance systems are provided with seals to ensure that nuclear material has not been introduced into or removed from it, or that data signals have not been tampered with. Seals do not provide physical protection but are designed to give evidence of possible manipulations.

Seals can be passive or active. An example of a passive seal is the conventional loop seal shown in Figure 2.11, which needs a wire loop to ensure continued closure. There are passive metal seals (like the one in Figure 2.11), glass seals, and ceramic seals. Other examples of passive seals are more high-tech seals such as the Cobra seal—a reusable fiber-optic loop seal with reflective particles in the seal body for tamper indication— and the ultrasonic sealing bolt for closing and securing containers of spent fuel assemblies underwater [60].

Active seals are equipped with remote monitoring capability. They continuously monitor seal integrity and record opening, closing, and tampering events. Examples are the T-1 radio frequency seal with

Figure 2.11. A seal deployed on a bolt locking the spent fuel pond in the Mochovce Nuclear Power Plant (photo courtesy of IAEA under CC BY-SA 2.0).

two-way RF communication, and the fiber optic loop sealing array RMSA with encrypted wireless transmission [61].

The IAEA also has the authority to use all information available to it in assessing whether a State has declared all nuclear material required to be safeguarded. According to comprehensive safeguards agreements and additional protocols, these include:

- information collected in the course of routine safeguards activities (e.g., inspections, accounting, surveillance, unattended and remote monitoring, etc.),
- information publicly available (commercial satellite imagery, open-source, and trade information), and
- information obtained from third parties.

Processing such a large amount of information requires advanced analytical techniques like artificial intelligence, i.e., the recreation by computers of human knowledge forms—and machine learning, i.e., programming based on data learning suitable for media, imaging, sensor data, etcetera.

References

[1] Smyth, H. D. 1945. Atomic Energy for Military Purpose. Maple Press, York.

[2] World Nuclear Association. 2021. Nuclear-Powered Ships. World Nuclear Association, November 2021. [Online]. Available: https://www.nytimes.com/1977/08/18/archives/soviet-nuclear-ship-reaches-north-pole-through-arctic-ice.html. [Accessed 26 February 2022].

[3] Shabad, T. 1977. Soviet Nuclear Ship Reaches North Pole Through Arctic Ice. The New York Times, pp. https://www.nytimes.com/1977/08/18/archives/soviet-nuclear-ship-reaches-north-pole-through-arctic-ice.html, 18 August 1977.

[4] US Department of Energy; Energy.gov. 2019. 9 Notable Facts About the World's First Nuclear Power Plant - EBR-I. Office of Nuclear Energy, 18 June 2019. [Online]. Available: https://www.energy.gov/ne/articles/9-notable-facts-about-worlds-first-nuclear-power-plant-ebr-i. [Accessed 26 February 2022].

[5] Nuclear Engineering International. 2004. Obninsk: number one. Progressive Media International, 13 July 2004. [Online]. Available: https://www.neimagazine.com/features/featureobninsk-number-one. [Accessed 26 February 2022].

[6] Argonne National Laboratory; Nuclear Engineering Division. 1955. Reactors Designed by Argonne National Laboratory; AEC Press release for BORAX-III lighting Arco, Idaho. Argonne National Laboratory, 12 August 1955. [Online]. Available: https://www.ne.anl.gov/About/reactors/borax3/index.shtml. [Accessed 6 April 2022].

[7] Haroldsen, R. 2008. The Story of the Borax Nuclear Reactor and the EBR-I Meltdown, Idaho Falls: Argonne National Laboratory; Nuclear Engineering Division.

[8] Hill, N. 2013. An Atomic Empire. A Technical History of the Rise and Fall of the British Atomic Energy Programme, https://doi.org/10.1142/p890: Imperial College Press.

[9] Zeng, E. 2018. Calder Hall: The First Industrial Nuclear Power Plant. Stanford University, http://large.stanford.edu/courses/2018/ph241/zeng2/.

[10] World Nuclear Association. 2022. Nuclear Power in the United Kingdom. World Nuclear Association, February 2022. [Online]. Available: https://world-nuclear.org/

information-library/country-profiles/countries-t-z/united-kingdom.aspx. [Accessed 26 February 2022].

[11] Craddock, J. 2016. The Shippingport Atomic Power Station. Stanford University, 13 March 2016. [Online]. Available: http://large.stanford.edu/courses/2016/ph241/craddock1/. [Accessed 26 February 2022].

[12] Fleger, P. A., I. H. Mandil and P. N. Ross. 1961. Shippingport Atomic Power Station Operating Experience, Developments, and Future Plans. In US-Japan Atomic Industrial Forum, Tokyo.

[13] Nuclear Newswire. 2017. Shippingport Atomic Power Station: Five Fast Facts. American Nuclear Society, 2 December 2017. [Online]. Available: https://www.ans.org/news/article-2004/shippingport-atomic-power-station-five-fast-facts/. [Accessed 26 February 2022].

[14] Wikipedia. 2022. Dresden Generating Station. Wikimedia, 22 February 2022. [Online]. Available: https://en.wikipedia.org/wiki/Dresden_Generating_Station. [Accessed 26 February 2022].

[15] Atomic Heritage Foundation. 2014. Soviet Atomic Program - 1946. Atomic Heritage Foundation, 5 June 2014. [Online]. Available: https://www.atomicheritage.org/history/soviet-atomic-program-1946. [Accessed 28 February 2022].

[16] Atomic Heritage Foundation. 2017. British Nuclear Program. Atomic Heritage Foundation, 16 March 2017. [Online]. Available: https://www.atomicheritage.org/history/british-nuclear-program. [Accessed 28 February 2022].

[17] Atomicarchive.com. 2022. The Soviet Response. Atomicarchive.com, [Online]. Available: https://www.atomicarchive.com/history/cold-war/page-7.html. [Accessed 1 March 2022].

[18] Atomic Heritage Foundation. 2017. French Nuclear Program. Atomic Heritage Foundation, 14 February 2017. [Online]. Available: https://www.atomicheritage.org/history/french-nuclear-program. [Accessed 28 February 2022].

[19] Zhang, H. and Y. Bai. 2015. China's Access to Uranium Resources. Harvard University, Cambridge.

[20] Atomic Heritage Foundation. 2018. Chinese Nuclear Program. Atomic Heritage Foundation, 19 July 2018. [Online]. Available: https://www.atomicheritage.org/history/chinese-nuclear-program. [Accessed 1 March 2022].

[21] Ban Treaty Organization (CTBTO). 2012. Operation 569 on 16 October 1964: China's First Nuclear Test. Ban Treaty Organization (CTBTO), 2012. [Online]. Available: https://www.ctbto.org/specials/testing-times/16-october-1964-first-chinese-nuclear-test#:~:text=The%20first%20Chinese%20test%20had,Soviet%20Union%2C%20Britain%20and%20France.. [Accessed 1 March 2022].

[22] International Atomic Energy Agency. 1953. Atoms for Peace Speech. International Atomic Energy Agency, IAEA, December 1953. [Online]. Available: https://www.iaea.org/about/history/atoms-for-peace-speech. [Accessed 2 March 2022].

[23] Goldschmidt, B. 1977. The origins of the international atomic energy agency. International Atomic Energy Agency Bulletin 19(4).

[24] International Atomic Energy Agency. 2022. History. IAEA, International Atomic Energy Agency, [Online]. Available: https://www.iaea.org/about/overview/history. [Accessed 1 March 2022].

[25] International Atomic Energy Agency. 2021. Operating Experience with Nuclear Power Stations in Member States 2021 Edition. IAEA, Vienna.

[26] International Atomic Energy Agency. 2016. IAEA, IAEA at a Glance: atoms for peace & development, Vienna: IAEA Office of Public Information and Communication.

[27] Carlson, J., V. Kuchinov and T. Shea. 2020. The IAEA's Safeguards System as the Non-Proliferation Treaty's Verification Mechanism. Nuclear Threat Initiative (NTI), Washington.

[28] International Atomic Energy Agency. 2021. IAEA Annual Report 2020, GC(65)/5. IAEA, Vienna.

[29] Watanabe, T. 1986. Multilateral conventions: Instruments of Global Cooperation. IAEA Bulletin 28(4).

[30] International Atomic Energy Agency. 1986. Convention on Early Notification of a Nuclear Accident, Vienna: IAEA.

[31] International Atomic Energy Agency. 1986. Convention on Assistance in the Case of a Nuclear Accident or Radiological Emergency, Vienna: IAEA.

[32] International Atomic Energy Agency. 1989. Statute as Amended up to 28 December 1989, Vienna: IAEA.

[33] U.S. Government Publishing Office. United States Code, 2014 Edition, Title 42 - The Public Health and Welfare, Chapter 23 - Development and Control of Atomic Energy, Washington: www.gpo.gov, 2014 Edition (of 1954).

[34] US NRC. 2021. Nuclear Materials. United States Nuclear Regulatory Commission, 04 November 2021. [Online]. Available: https://www.nrc.gov/materials.html. [Accessed 5 March 2022].

[35] EUR-Lex Access to European Union Law. 2012. Consolidated Version of the Treaty Establishing the Europen Atomic Energy Community, Document 12012A/TXT. Euratom, 26 October 2012. [Online]. Available: https://eur-lex.europa.eu/legal-content/ EN/TXT/?uri=CELEX%3A12012A%2FTXT. [Accessed 10 March 2022].

[36] International Atomic Energy Agency. 1970. INFCIRC/140, Treaty on the Non-Proliferation of Nuclear Weapons. IAEA, London, Moscow and Washington.

[37] International Atomic Energy Agency. 1972. INFCIRC/153 (Corrected), The Structure and Content of Agreements Between the Agency and States Required in Connection with the Treaty on the Non-Proliferation. IAEA, Vienna.

[38] International Atomic Energy Agency. 1968. INFIRC/66/Rev.2, The Agency's Safeguard System (1965, as Provisionally Extended in 1966 and 1968). IAEA, Vienna.

[39] International Atomic Energy Agency. 1997. INFCIRC/540 (Corrected), Model Protocol Additional to The Agreement(s) Between State(s) and the International Atomic Energy Agency for the Application of Safeguards. IAEA, Vienna.

[40] Johnson, L. 1998. Material Obligations: Forms and Content. In Technical Workshop on Safeguards, Verification Technologies, and Other Related Experience, Vienna.

[41] Editor International Atomic Energy Agency. 2022. Nuclear Law. The Global Debate, The Hague: TMC Asser Press.

[42] Bytchkov, V. and J. N. Cooley. 2020. IAEA Safeguards System: Implementing the State-Level Concept. NTI, Center for Energy and Security Studies, Washington.

[43] Arno, M. 2018. LLNL-TR-745003, Standardizing Acquisition Path Analysis: Quantifying a State's Ability to Establish and Clandestinely Operate a Nuclear Facility. Lawrence Livermore National Laboratory, Livermore.

[44] Liu, Z. and S. Morsy. 2001. Development of the Physical Model. In Symposium on International Safeguards: Verification and Nuclear Material Security, Vienna.

[45] International Atomic Energy Agency. 2008. Nuclear Material Accounting Handbook, IAEA Service Series 15, Vienna: IAEA.

[46] International Atomic Energy. 2004. Regulatory Control of Radiation Sources, Safety Guide No. GS-G-1.5 Jointly sponsored by FAO, ILO, PAHO and WHO, Vienna: IAEA.

[47] U.S. NRC. 2022. Office of Nuclear Material Safety and Safeguards. U.S. Nuclear Regulatory Commission, 4 March 2022. [Online]. Available: https://www.nrc.gov/ about-nrc/organization/nmssfuncdesc.html. [Accessed 27 March 2022].

[48] U.S. NRC. 2020. 10 CFR PART 75—Safeguards on Nuclear Material—Implementation of Safeguards Agreements Between the United States and the International Atomic Energy Agency. U.S. Nuclear Regulatory Commission, 25 November 2020. [Online].

Available: https://www.nrc.gov/reading-rm/doc-collections/cfr/part075/full-text.html. [Accessed 27 March 2022].

[49] U.S. NRC. 2020. PART 74—Material Control and Accounting of Special Nuclear Material. U.S. Nuclear Regulatory Commission, 23 September 2020. [Online]. Available: https://www.nrc.gov/reading-rm/doc-collections/cfr/part074/index.html. [Accessed 27 March 2022].

[50] Oakberg, J. and K. Gilligan. 2014. ORNL/TM-2014/404, Nuclear Material Accounting and Reporting Information Systems: Capabilities Review. Oak Ridge National Laboratory, Oak Ridge.

[51] NAC International. 2022. NAC Reporter™. NAC International, 2022. [Online]. Available: https://www.nacintl.com/solutions/consulting-services/material-control-and-accountability/nac-reporter. [Accessed 2 April 2022].

[52] NNSA Office of Nonproliferation and Arms Control (NPAC). 2021. Introduction to International Safeguards, Washington: US Department of Energy.

[53] Addleman, R., W. Chouyyok, B. Naes, D. Willingham, B. McNamara, A. Spigner and K. Olsen. 2016. PNNL-26042, Materials and Methods for Streamlined Laboratory Analysis of Environmental Samples, FY 2016 Report. Pacific Northwest National Laboratory, Richland.

[54] Yeongduk, K. 2016. Detection of antineutrinos for reactor monitoring. Nuclear Engineering and Technology 48: 285–292.

[55] Borstad, G. A., J. Lim, L. N. Brown, Q. B. Truong et al. 2004. Satellite hyperspectral imaging in support of nuclear safeguards monitoring. In Proceedings of the 45th Annual Meeting of the Institute of Nuclear Materials Management, Orlando.

[56] International Atomic Energy Agency. 2014. IAEA Safeguards Overview: Comprehensive Safeguards Agreements and Additional Protocols. IAEA, 2014. [Online]. Available: https://www.iaea.org/publications/factsheets/iaea-safeguards-overview. [Accessed 13 March 2022].

[57] Bunn, G. 2007. Nuclear Safeguards. How far can Inspectors go? IAEA Bulletin 48(2), March 2007.

[58] International Atomic Energy Agency. 2011. INV Series No. 1 (Rev. 2), Safeguards Techniques and Equipment: 2011 Edition, Vienna: IAEA.

[59] International Atomic Energy Agency. 2001. IAEA/NVS/3, IAEA Safeguards Glossary 2001 Edition, Vienna: IAEA.

[60] International Atomic Energy Agency. 2016. Surveying Safeguarded Material 24/7. IAEA, 12 September 2016. [Online]. Available: https://www.iaea.org/newscenter/news/surveying-safeguarded-material-24/7. [Accessed 11 October 2023].

[61] Sandia National Laboratories. 2013. SAND2013-8685C, Containment and Surveillance. An Overview, Albuquerque: International Safeguards and Technical Systems Department. SNL.

CHAPTER 3
Nuclear Fuel Cycle Overview

The fuel used in the majority of current nuclear reactors is uranium in the form of uranium oxide (UO_2), natural or enriched in ^{235}U. Therefore, natural uranium requires a series of processes to change into the right chemical form, purify it, and increase its ^{235}U content. Light water reactors (LWR) and other commercial reactors use low-enriched uranium (LEU) with up to 5% ^{235}U. Highly enriched uranium (HEU) with $\geq$ 20% ^{235}U is mostly used in research reactors and nuclear weapons [1]. These processes, together with methods to manage and dispose of nuclear waste, including spent fuel, are known as the nuclear fuel cycle. The main stages of the fuel cycle are shown in Figure 3.1.

The nuclear cycle starts with the mining and milling of the uranium ore. Next, the uranium is purified and converted into UF_6 to facilitate its enrichment in ^{235}U. Enriched uranium in the form of UO_2 then goes to fuel rod fabrication. Finally, the fuel is burned in the nuclear reactor to produce electricity.

After burnup, the spent fuel is first stored in pools at the reactor site for cooling. Afterward, in an open fuel cycle, it can be kept in dry cask storage systems waiting for final disposal. Otherwise, in a closed cycle, the fuel is sent for reprocessing. Recovered U and Pu (plutonium) from reprocessing is used for mixed oxide fuel (MOX) fabrication. This is a fuel made of depleted uranium dioxide (UO_2) and recycled plutonium dioxide (PuO_2). Nuclear waste from reprocessing is also stored interim waiting for final disposal. A deep geological repository is intended for the disposal of both nuclear waste and spent fuel.

Figure 3.1 includes two more possible processes in progress. They are the deconversion of depleted uranium hexafluoride (DUF_6) tails from enrichment to obtain a mix of depleted UO_2 and U_3O_8 (triuranium octoxide) suitable for disposal as low-level radioactive waste (LLW) at a licensed disposal facility [2], and the re-enrichment of part of the DUF_6 tails using a third-generation laser process to further produce high-assay

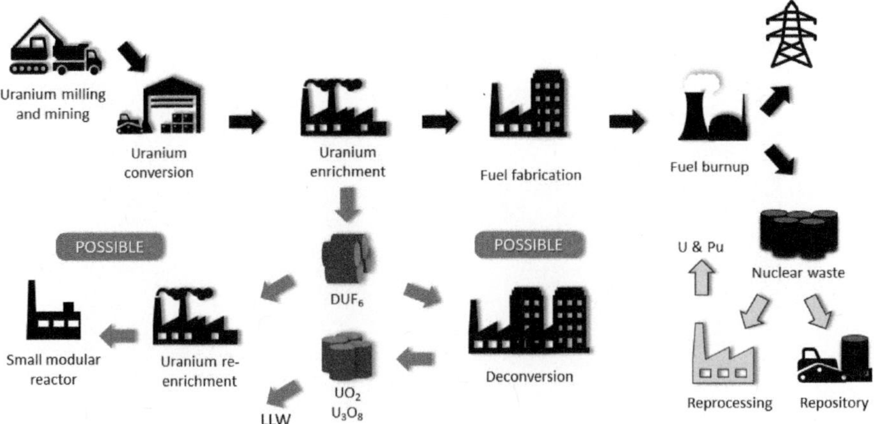

Figure 3.1. The nuclear fuel cycle.

low-enrichment uranium (HALEU) for new generation small modular reactors [1].

Uranium mining and milling

Uranium is found in very low concentrations almost everywhere—in soil, rocks, and water—and is the only naturally occurring fissile element on Earth. Currently, the most important producers of uranium are Kazakhstan, Canada, and Australia, which account for about 64% of world production [3].

Because uranium ore is just a rock with elevated amounts of uranium in it, deposits can occur in any type of rock, i.e., sedimentary, metamorphic, and igneous. Consequently, conventional milling[1] is often necessary to retrieve the uranium from the extracted ore.

There are two methods for extracting uranium ore from deposits: conventional mining, and *in situ* leach mining. Conventional mining includes open-pit mining, used when the deposits are close to the surface, generally less than 100 meters deep; and underground mining, used for deeper deposits and higher ore grades [4, 5].

Conventional open-pit mining strips away the earth and rock that lie above the uranium via drilling or blasting. Large earthmoving equipment, e.g., excavators, backhoe loaders, dump trucks, etc., are used to break the ore, strip waste material, and load them into trucks or conveyors for transport. Unwanted soil and rock are typically put aside for future reclamation.[2]

---

[1] Ore milling is the process of separating commercial valuable materials from ore.

[2] Reclamation is the process of restoring disturbed areas to a previous natural resource setting, such as forest or agricultural land uses while minimizing environmental impacts.

Figure 3.2. Aerial view of the open-pit Rössing uranium mine in Namibia (photo courtesy of Ikiwaner under license Wiki Commons).

To reach greater depth, the site is typically excavated with a series of benches that are between four and sixty meters in size, depending on the machinery used to excavate [6]. A road forming a ramp-up is usually dug at the side of the pit for ore-carrying trucks to transport the rock to the milling facility. These benches can be observed in Figure 3.2. Figure 3.3 shows how the walls are dug at an angle to prevent landslide occurrences. Additional ground support may also be required.

Figure 3.3. An excavator working at an open-pit uranium mine.

An open-pit uranium mine includes several facilities, e.g., water treatment and distribution systems, storage areas for ore and wastes, and a few infrastructure buildings including a space for changing clothes for employees, equipment maintenance, etc. Waste rock[3] is usually piled up near the edge of the open pit in a waste dump that is also tiered and stepped to prevent landslides.

The conventional underground mining method is used to mine higher-grade deposits (up to 20% uranium) located more than 100 meters below the surface. Access to underground deposits is achieved by digging vertical shafts or downward slopes to the depth of the ore. A series of horizontal tunnels, ramps, and chambers are then built to access the ore [7].

Because of the likelihood of exposure to radionuclides from uranium radioactive decay, particularly ^{226}Ra (radium 226) and its decay product ^{222}Rn (radon 222), underground mining can pose a radiation hazard to miners in addition to the other risks linked to the mining sector. Adequate ventilation systems to continuously replace the air contaminated with ^{222}Rn, planned radiation monitoring of workers and workplaces, the use of personal respirators, and when necessary, shotcrete on walls and shielding in loading and hauling vehicles to reduce the dose from γ radiation, are some of the radiation safety measures that should be taken at the mine [8]. To protect the public, vented ^{222}Rn gas must not exceed the limits established by the national regulatory authority [9].

At some deposits, the surrounding water could be a major issue too. For the most part, it is because the ^{222}Rn dissolved in the groundwater

---

[3] Waste rock is the bedrock that has been mined and transported out of the pit but does not have metal concentrations of economic interest.

under pressure can be released into the mine when the water loses its pressure. In such events, radiation levels can be extremely high. To remove the heat and freeze the water within the soil/rock pore spaces, the ore body is cooled with calcium chloride ($CaCl_2$) rich brine at –40°C. This protects from radon emissions, prevents water inflow, and stabilizes ground conditions. The chilled brine from a refrigeration plant is circulated within large pipes through the ground surrounding the ore in a linear or grid pattern for about a year [10].

Most of today's mining practices do not require workers to be within the close limits of underground excavations. Current design layouts include non-entry type mining methods such as raise bore, boxhole boring, blast hole stoping, and jet boring. Besides, the mining and hauling equipment to collect the ore and clean the floor from the waste rock is remotely controlled and operated from areas far from the orebody [7, 11, 12]. The selected mining method depends on a range of factors, including the quality and quantity of the ore, the depth, and shape of the orebody, and so on. A brief explanation is presented next.

In raise bore[4] mining, a horizontal underground shaft is made below the uranium deposit to collect with remote-controlled scooptrams the ore loosed by a raising-up reaming head. The cuttings fall by gravity when the reaming head is pulled back toward the drilling machine located in the chamber above the orebody (See Figure 3.4). When the dug hole is complete, it is filled with concrete, which helps further stabilize the mine.

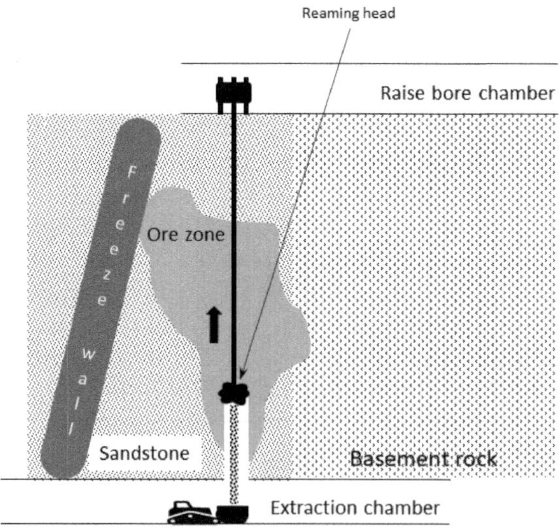

Figure 3.4. Raise bore mining method scheme.

---

[4] A raise borer is a drilling machine to excavate between two levels of a mine without the use of explosives.

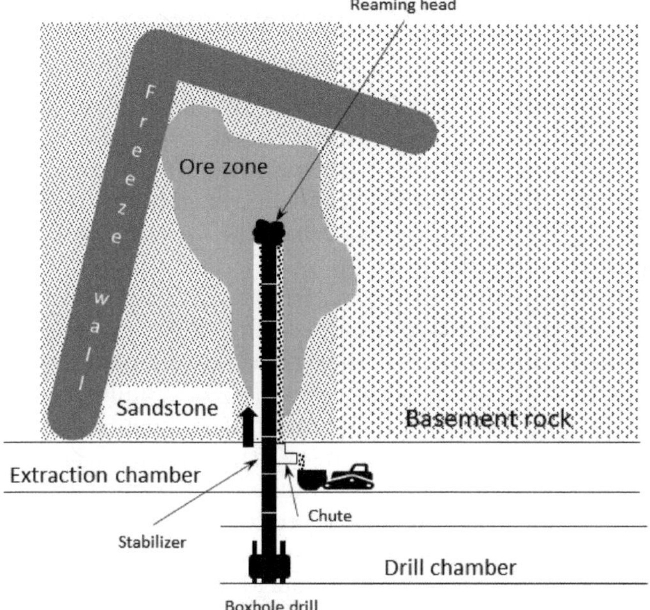

Figure 3.5. Boxhole mining method scheme.

The boxhole boring is similar to the raise bore method, but the drilling machine is located below the mineralization. A series of holes are drilled upward through the ore using the revolving reaming head from a drill machine located in a chamber made in the waste rock below the ore zone. At fixed intervals, several stabilizers are used to keep the drill rods centered while raising the reaming head (See Figure 3.5). The falling ore is collected from a chute in the extraction chamber.

Blast hole stoping[5] involves drilling access above the mineralization, and extraction access below it. Blast holes are drilled in a fan-like pattern, in the stope, i.e., the area between the upper and lower access levels. They are next loaded with explosives and detonated. The broken ore is removed by remote-controlled load-haul dump machines or rail cars.

In jet boring mining, a boring nozzle is used to cut out cavities in the ore zone using water under high pressure. A production tunnel is first excavated and encased with concrete across the frozen rock to place the jet boring system. A pilot hole is next drilled into the orebody to insert the jet that will do the cutting. The resulting ore slurry is collected through a

---

[5] Stoping is the opening of large underground rooms, or stopes, by the excavation of ore.

network of pipes and sent to the underground grinding and thickening circuits [13].

Underground grinding and thickening circuits are generally beneficial to prepare the uranium ore for transport. They use a series of remote-controlled chutes to send the coarse ore to be crushed to the consistency of fine sand, mixed with water to form a slurry, and pumped to a storage tank located in a surface loadout facility. At regular intervals, the slurry is loaded into transport containers Type IP-2[6] arriving by truck to be sent to the milling facility.

In-situ leaching (ILS), also called *in-situ* recovery (ISR), the second method for extracting uranium ore from deposits, is an alternative underground mining process that involves minimal environmental disruption and does not produce waste rock or tailings. This method involves extracting the uranium by injecting lixiviants into the orebody through injection wells. The pregnant leach solution, i.e., the solution containing the dissolved uranium is taken out using production wells. This leaching method is especially suited for deposits typically associated with relatively shallow aquifers, with sufficient permeability [14].

Most conventional lixiviants are acid and alkaline solutions. The selection of one or the other depends on the carbonate content of the rock that is hosting the uranium orebody, which is usually > 2% CO_2 for alkaline lixiviants. Different oxidants, e.g., H_2O_2, oxygen, and air, are added to increase the efficiency of leaching. ISL method has been steadily increasing, and in 2022, accounted for the largest production category in the world [15].

After mining, uranium ores or slurries are sent to the milling facility to produce uranium ore concentrate (UOC). This is the solid form historically known as yellowcake commonly referred to as U_3O_8.[7] The leaching processes to produce uranium ore concentrate from mined ore are illustrated in Figure 3.6.

The particular process design to use depends on the ore deposit mineralogy, gangue material, oxidation state, and contaminants present. Similar to *in situ* leaching, alkaline leach is recommended for the treatment of acid-consuming minerals. Also, when the uranium concentration is high enough, solvent extraction is typically selected over ion exchange in the purification process because of cost reasons.

---

[6] According to the regulations for the safe transport of radioactive materials, an industrial package specially designed for low specific activity material.

[7] Most modern uranium recovery facilities produce a yellowish compound of uranyl peroxide dihydrate [$(UO_4 \cdot nH_2O)$].

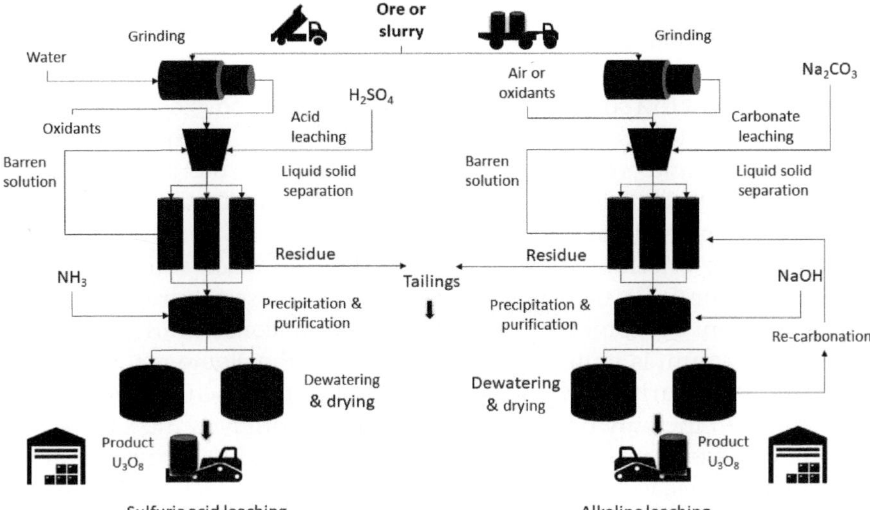

Figure 3.6. Uranium ore concentrate manufacturing acid and alkaline processes.

The production of uranium concentrate generally involves the following operations [16]:

- Ore blending and grinding (ore or slurry)
- Extraction of uranium from the ore by leaching
- Decantation of solids from the pregnant leach solution
- Removal of impurities from the uranium solution by solvent extraction or ion-exchange
- Precipitation of uranium from the purified solution
- Dewatering, drying, and calcining of the ore concentrate.

Upon arriving at the milling facility, coarse uranium ore and ore slurries are sampled and weighted to establish the moisture content and accordingly adjust the blending and crushing operations. Next, the ore is mixed with water and sent to the mills to be blended and crushed to the appropriate size for leaching. Repeated grindings using ball mills, autogenous mills, and/or pebble mills may be required to obtain the required size. The ore exits the mills in the form of a slurry containing 70% solids and 30% water.

The ore slurry is next pumped to the leaching agitators where, depending on the leaching process, sulfuric acid (H_2SO_4) or sodium carbonate (Na_2CO_3) is added to dissolve the uranium and separate it from other minerals present in the slurry. Heat and oxidants such as O_2, hydrogen peroxide (H_2O_2), manganese dioxide (MnO_2), air, or sodium bicarbonate ($NaHCO_3$) can be added to expedite the uranium dissolution and improve its recovery. Oxidants help to convert less soluble U^{4+} in the ore to more soluble U^{6+} [16].

After the decantation of solids, the pregnant leach solution, rich in uranium, is sent to a solvent extraction process or an ion exchange process. The purpose is to free the uranium from other metals such as vanadium, molybdenum, and iron that can also be carried by the pregnant leach solution. The leach solution is also filtered before extraction to eliminate any remaining solid particles of sludge [9].

When the uranium concentration in the acid leach solution is high (> 800–900 mg/l U) the solvent extraction process is preferred. Both processes can be combined for low uranium concentrations. In this case, the uranium is eluted[8] from the resin and the eluate is then further processed and purified by solvent extraction. If the leaching process was alkaline, then the only option is ion exchange. Until now, no satisfactory solvent extraction process seems to have been found to recover uranium from carbonate leach solutions [17].

A series of countercurrent mixers/settlers are used in solvent extraction to put in contact the leach solution with a kerosene solution of tertiary amines[9] and a diluent modifier[10] such as isodecanol. The extraction of uranium is achieved by advancing the solvent through the circuit.

In the ion exchange process, the leach solution or pulp is passed through columns filled with organic, synthetic resins, containing anionic functional groups capable of exchanging for the anionic uranyl sulfate or carbonate uranyl complexes in the solution. The process comprises a series of adsorption, elution, and resin regeneration steps. When the resin is filled with uranyl ions the column is deemed loaded and ready for elution. Next, the resin is eluted with a concentrated chloride salt solution to obtain a purified uranium eluate. The regeneration step discharges the resin from strongly adsorbed ions that compete with uranyl sulfate anions and compounds.

Concentrates from solvent extraction and pregnant elute solutions from ion exchange are next directed to the precipitation circuit and the barren leach solutions are recycled back to the leaching circuit.

Slurries from liquid-solid separation are thickened and pumped to tailings storage facilities. Tailings may be mined out like open pits, used for backfilling underground mine openings, or disposed of in specially designed dams [18].

---

[8] Elute means to remove an adsorbed substance by washing it with a solvent. The resulting solution is the eluate.

[9] Molecules that contain three C-N (carbon-nitrogen) bonds.

[10] A high-molecular-weight alcohol that helps phase disengagement.

Uranium is afterward precipitated from either the pregnant solvent solution or the pregnant elute solution neutralizing it with ammonia hydroxide, sodium hydroxide, or lime depending upon whether the leaching process was acid or alkaline. H_2O_2 may also be added. If the leaching process was alkaline, the carbonate ions must be removed before the precipitation to lower the pH. A wet concentrate of ammonium diuranate (ADU) is obtained if the process is acid or sodium diuranate (SDU) if alkaline. Uranium peroxide (UO_4) is precipitated if H_2O_2 is used.

The resulting precipitate is washed to remove residues, filtered, and dried in a series of thickeners, filters, and/or centrifuges. ADU is further calcined in furnaces to obtain oxides of uranium, most commonly U_3O_8. During calcination, yellow ADU transitions from orange uranium trioxide (UO_3) to a green-grayish U_3O_8, and UO_4 transitions to U_3O_8 with a hydride of $UO_3.5 \cdot H_2O$ as an intermediate compound.

On the other hand, sodium diuranate (SDU) impurities (especially Fe and rare earth elements) affect the specifications of the uranyl nitrate to be converted into UF_6 or UO_2. Therefore, it also has to be transformed into ammonium diuranate or uranium peroxide [19, 20, 21].

Following the regulations for the safe transport of radioactive materials [22], the calcined uranium concentrate, U_3O_8, is then packaged in an industrial package Type 1 (IP-1). Typically, this involves packaging the concentrate in a steel drum, which has a removable lid and a bolted retaining ring, similar to the one shown in Figure 3.7. The nominal volume of the drum is 210 liters. It is recommended that these drums be suitable

Figure 3.7. Industrial package (IP-1) for radioactive material of low specific activity (LSA) (Photo courtesy of Kazatomprom under Creative Commons license).

for repeated handling, stacking, and storage for extended periods. This is because some drums may remain in the storage area at the converter for up to two years before processing.

U_3O_8 drums are to be labeled as dangerous goods Class 7 and assigned the number of the United Nations (UN) 2912—radioactive material of low specific activity (LSA-I), non-fissile, or fissile-excepted. They are generally shipped to the conversion facility by truck.

Uranium conversion

Yellowcake (U_3O_8) is a stable compound, suitable for shipping between processes, but not for enrichment. It is because the process of enriching requires uranium in the form of uranium hexafluoride (UF_6), a compound that can be a solid, liquid, or gas within a reasonably small range of temperature. Likewise, the uranium concentrate still contains impurities that must be further removed. The purification varies according to the method selected for the conversion of U_3O_8 into UF_6, which can be dry [23] or wet [24].

Figure 3.8 attempts to represent the steps of the two approaches for the conversion process. The first method uses a dry, fluoride volatility path to directly convert the U_3O_8 into UF_6 and purify the fluoride via fractional distillation at the end of the process. This path is distinguished in Figure 3.8 using dashed lines. The second method uses acid digestion first and then, a subsequent solvent extraction to purify the U_3O_8 before its conversion into UF_6.

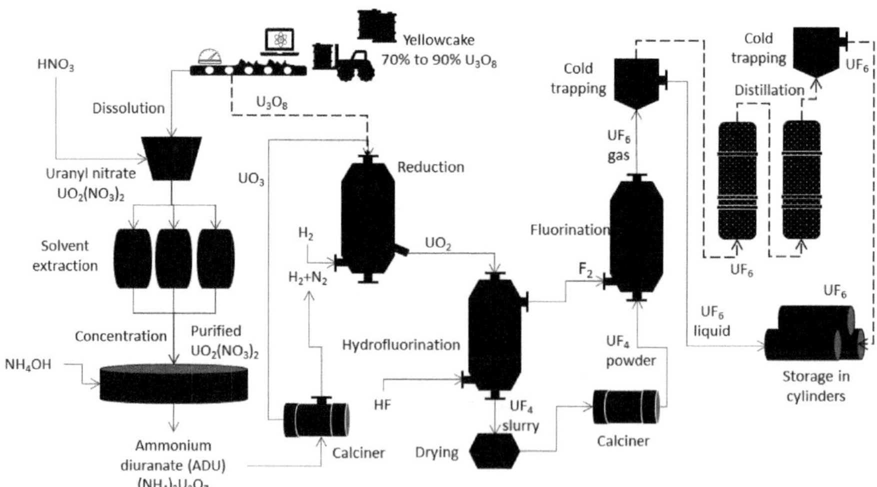

Figure 3.8.　UF_6 conversion process.

It is important to note that only four conversion plants are currently operating in the world. They are Pierrelatte & Malvési, in France, operated by Orano S.A.; Lanzhou & Hengyang, in China, operated by China National Nuclear Corporation (CNNC); Port Hope, in Canada, operated by Cameco Corporation; and Seversk, in Russia, operated by Rosatom [24]. The sole plant in the U.S. and the only one with a dry process, Honeywell's Metropolis Works, in Metropolis, Illinois, was shut down at the end of 2017. It was planning to restart production in 2023 but this is uncertain [25].

The U_3O_8 from milling arrives at the conversion facility in the steel drums described before and is stored for processing. The first step is weighing the material and analyzing the uranium concentration to ensure it meets the specifications. Calibrated scales are used to determine the net weight of the product in each drum. Gamma spectroscopy, X-ray fluorescence, and Auger electron spectroscopy can be used to verify the uranium contents with the necessary accuracy [26].

U_3O_8 typically contains 70–90% percent uranium and up to 20% impurities that have to be removed. In the wet process, the U_3O_8 from milling is diluted with nitric acid (HNO_3) to produce uranyl nitrate $[UO_2(NO_3)_2]$ that is afterward fed to countercurrent solvent extraction columns for purification. The solvent extraction process uses tributyl phosphate dissolved in kerosene or dodecane as an extractant. Purified uranyl nitrate is washed from the extractant with water. This solution generally contains very few contaminants and is further concentrated and injected with NH_4OH (ammonium hydroxide) to precipitate ammonium diuranate [27]. As shown in Figure 3.8, the ammonium diuranate is further calcined at 600°C to obtain pure UO_3. Dissociated NH_3 from the calciner supplies an extra mixture of hydrogen and nitrogen to the fluidized bed reactor, granting a 99% UO_3 reduction to UO_2 [28]

Although not shown in Figure 3.8, in the dry process, the U_3O_8 is first calcined to remove some impurities, crushed, and passed through meshes to the desired size, to be reduced to UO_2 with H_2 in the fluidized bed reactor.

In both the dry and the wet processes, uranium tetrafluoride (UF_4), or green salt, is an intermediate compound produced when reacting UO_2 with an excess of anhydrous hydrogen fluoride (HF). This reaction is called hydrofluorination and is done in a series of countercurrent Monel[11] fluid beds.[12] Yielded UF_4 slurry from hydrofluorination is then dried and calcined to powder before entering the fluid bed fluorination reactors.

---

[11] Monel is a strong rust-resistant stretchable nickel-copper alloy. Other similar materials are Hastelloy and Inconel.

[12] It is worth noting that in the mid-1980s, all fluidized beds for UO_2 and UF_4 production in France and the UK were replaced by kilns, which significantly benefited safety.

The UF_4 will react with elemental fluorine gas (F_2) in the fluid bed reactors to form gaseous uranium hexafluoride (UF_6). The reaction occurs at temperatures above 482°C and is highly exothermic. To act as a heat dissipator and collect residual uranium and nonvolatile uranium progeny products, a bed of calcium fluoride (CaF_2) is used as inert material [23].

The F_2 gas spent in fluorination is usually prepared on-site using electrolytic cells to dissociate the HF [29]. Before entering the cold traps in the next step, the UF_6 gas is passed through filters to remove any solid particles left. Various cold traps, operating at –30°C and below, cool the gaseous UF_6 into a white crystalline solid [30]. When the cold traps are full, they are heated to liquefy the crystallized UF_6. A portion of cold trap off-gases is recycled to the fluorination reactors, and the remainder is treated for the removal of fluorides and residual uranium before atmospheric discharge.

In the wet process, the product coming from the cold traps is at least 99.95% pure. In the dry process, however, the UF_6 from the cold traps contains impurities (e.g., fluorides of tungsten, molybdenum, thorium, protactinium, sulfur, carbon, etc.) that need to be removed before it can be used for enrichment. Therefore, the final step for the dry process is a fractional distillation consisting of two columns, one to remove the fluorides boiling at a higher temperature than UF_6 and the other, to remove the low boilers such as WF_6 or MoF_6 (tungsten hexafluoride or molybdenum hexafluoride) [31].

To perform refining, gaseous UF_6 is fed to the first column. The concentrated low boiler fraction is volatilized and exits from the top of the column to be removed at an overhead condenser (not shown in Figure 3.8). Impurities not removed in the condenser are fed back to the system just before the cold trap. The UF_6 purified to 99.99% is removed at the top of the second column and goes to the cold trap. The bottom of this column contains the high boiling/non-volatile fluorides to be disposed of as waste. Among them, are thorium, and protactinium fluorides, which are gamma emitters and add radiation to the safety hazards.

The liquid UF_6 from the cold traps in both processes—dry and wet—is finally drained into approved, specially designed 14 t steel cylinders type 48Y, like the one shown in Figure 3.9, suitable for the UF6 storage and transport to the enrichment plant. After the cylinders are filled, leaving a 5% shortage, they are left to cool to ambient room temperature. This causes the UF_6 to become a white crystalline solid. During filling, the quantity and quality of the UF_6 are continuously measured.

Cylinders 48Y containing solid UF_6 are labeled as dangerous goods Class 7 with Class 8 (corrosive) as a subsidiary risk. The assigned United Nations number is 2978, meaning radioactive material, uranium hexafluoride, non-fissile or fissile excepted [22].

Figure 3.9. A 48Y cylinder for natural uranium hexafluoride storage and transport (photo courtesy of the US Department of Energy).

According to applicable standards and regulations, cylinders 48Y should withstand a hydraulic test at an internal pressure of at least 1.4 MPa; the free drop test; and a fire test at 800°C for 30 min, without leakage, loss or dispersal of the uranium hexafluoride [22, 32].

Uranium enrichment

Currently, only a fraction of all operational reactors uses natural uranium as fuel. The vast majority of them, approximately 95%, are light water-cooled reactors and require low-enriched uranium (LEU) as fuel [33]. The enrichment levels for these reactors typically range from 3.0 to 5.0% of ^{235}U. The process of enrichment, based on the difference of mass between the two uranium isotopes ^{235}U and ^{238}U in the natural ore, involves increasing the quantity of ^{235}U. This process uses the UF_6 in gaseous form obtained before.

During the Manhattan Project, various methods of enrichment were experimented with and industrialized. For instance, the electromagnetic and thermal processes were employed to obtain the enriched uranium needed for the bomb. However, these methods were abandoned shortly after the war due to their poor efficiency and enormous energy consumption. Since WWII, only two methods have been operated commercially: the gaseous diffusion process and the gas centrifuge process.

The first-generation enrichment technology, gaseous diffusion, was dominant during the 20th century despite its small separation factor. This method required vast amounts of electricity and extensive facilities, rendering it obsolete in the present day. The last diffusion enrichment facility in the US, the Paducah plant in Kentucky, which was one of the world's largest suppliers, was terminated in 2013 [34]. At its peak, this plant consumed 3,000 MW, enough energy to power a city of about half a million people. The other two plants in the US, located in Piketon, Ohio, and Oak Ridge, Tennessee, were deactivated much earlier [35].

In gaseous diffusion, the uranium isotopes ^{235}U and ^{238}U in the form of gaseous UF_6 are separated based on a slight difference in velocity. Considering that the rate of diffusion is inversely proportional to the square root of the mass and that the difference in weight between these two isotopes is slightly greater than 1%, the enrichment factor would be about 0.14% per stage [36]. Though, theoretically, to enrich ^{235}U from 0.7% to just 1.0% takes approximately 347 stages.

Each stage consisted of a diffuser with barrier tubing, a compressor, and a control valve. As illustrated in Figure 3.10, the UF_6 gas that was pumped into the barrier tubing moved through special filters or porous membranes with holes so small that only the lightest $^{234}UF_6$ and $^{235}UF_6$ gas molecules could pass through quickly enough. However, it took many hundreds of continuous barriers before the UF_6 gas in the enriched stream could get enough ^{235}U for fuel fabrication.

A second-generation technology, the gas centrifuge process, replaced gaseous diffusion. Large commercial enrichment plants using the gas centrifuge process are operating in France, Germany, the Netherlands, the UK, the USA, Japan, China, and Russia.

In the gas centrifuge process, the UF_6 gas from the conversion is fed into the centrifuge cylinder and rotated at a high speed to create a strong centrifugal force. The heavier gas molecules, $^{238}UF_6$, move towards the outside of the cylinder while the lighter gas molecules, $^{235}UF_6$, gather closer to the center. The slightly enriched ^{235}U stream is then withdrawn and fed into the next centrifuge, i.e., the next higher stage. The stream with a lower concentration of ^{235}U is recycled back into the next lower stage.

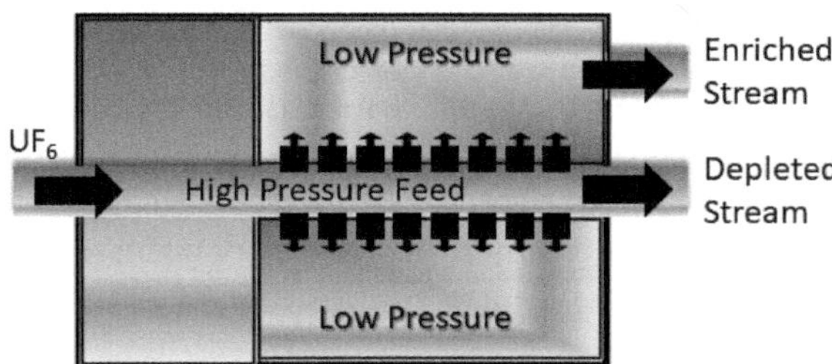

Figure 3.10. Simplified drawing of a gaseous diffusion stage.

Figure 3.11. Cascade of gas centrifuges used to produce enriched uranium (photo courtesy of the US Department of Energy).

This process continues in a cascade[13] until the desired $^{235}UF_6$ enrichment is achieved.

Figure 3.11 shows the cascade of centrifuges that were used for enrichment in Piketon, Ohio, in 1984. Today centrifuges are much smaller.

The US's only gas centrifuge commercial plant is the URENCO facility licensed to Louisiana Energy Services (LES) in Eunice, New Mexico [35]. Another plant, the American Centrifuge Plant (ACP) of Centrus Energy Corp., which was under construction in Piketon, Ohio just started operation in October this year 2023 [37]. The plant construction was authorized by the NRC in 2021 to demonstrate gas centrifuge technology for the commercial production of high-assay low-enriched uranium (HALEU). In partnership with the US Department of Energy and Oak Ridge National Laboratory, they developed a 16-centrifuge cascade to produce $^{235}UF_6$ with 19.75% ^{235}U [38]. Emerging advanced Small Modular Reactors require high-assay low-enriched uranium of about 20% ^{235}U enrichment.

A centrifuge can separate isotopes much more efficiently than a gaseous diffuser, and the process uses less energy, although several stages are still required, and the process is costly. Enrichment accounts for nearly 30–50% of the cost of nuclear fuel and approximately 5% of the total cost of the electricity generated by nuclear power [39].

---

[13] A cascade consists of a group of elements or separation units arranged in parallel to make up a single stage. They receive identical feed and generate the same product and waste.

Innovative technologies using lasers to separate uranium isotopes might provide much higher enrichment compared to the two previous methods. Potentially, they are offering significantly lower overall costs. Lasers are believed to selectively excite atoms or molecules containing ^{235}U so they can be preferentially extracted. The first two of these technologies appeared in the period 1970 to 1990 when high-intensity tunable lasers became available. One of them is called atomic vapor laser isotope separation (AVLIS) and the other is called molecular laser isotope separation (MLIS). Both are based on the selective excitation of the energy levels of ^{235}U and ^{238}U by radiation of a particular frequency [40].

Studies in laser technologies at the time included devices to feed uranium-metal vapor for selective photoionization (AVLIS) or to vaporize a uranium compound for photo-dissociation or chemical activation (MLIS); laser and optical systems to selectively excite ^{235}U species; and methods to collect the enriched product and tails. Such technologies offered lower energy inputs, lower capital costs, and lower tails assays. However, they were not competitive with the gas centrifuge because of material corrosion, low laser pulse repetition rates, and inefficient laser excitation, among other concerns [41].

A third-generation laser enrichment process employs a laser to vibrationally excite ^{235}UF$_6$ in a gas mix (^{235}UF$_6$:G) flowing through a low-pressure vessel and uses condensation repression for isotope separation. Such a process is not well known because detailed information is not yet publicly available. Nonetheless, two methods, known as CRISLA (Condensation Repression Isotope Separation by Laser Activation) and SILEX (Separation of Isotopes by Laser Excitation) are of interest. CRISLA was reported in a US Patent granted in 2019 [42] and SILEX, under commercial development, was created by Silex Systems Ltd., an Australian company now the owner of the majority of GLE (Global Laser Enrichment) in the US [43]. In 2012, GLE received an NRC license for constructing and operating a full-scale laser enrichment facility in Wilmington, North Carolina, where SILEX completed its technology validation at prototype scale in 2013 through detailed testing in a Test Loop[14] [44].

As part of the Paducah project, GLE has an agreement with the US Department of Energy to purchase about 300,000 tons of depleted uranium tails, stored in Paducah, Kentucky, in the form of DUF$_6$ (depleted uranium hexafluoride) with approximately 0.25% to 0.4% of ^{235}U to re-enrich them to natural grade uranium (~ 0.7% of ^{235}U) using SILEX technology. The Paducah project is still subject to several factors, including the satisfactory

---

[14] Test loop is a versatile experimental facility for the demonstration of the technology at the prototype level. It helps to find and identify any deficiencies that might be present.

completion of the engineering scale-up program mentioned before and uranium market conditions [45].

According to R. Snyder [41], the laser employed by SILEX to selectively excite the $^{235}UF_6$ is a 10 µm CO_2 TEA (Transversely Excited Atmospheric) laser with a liquid-nitrogen-cooled para hydrogen-filled Raman conversion cell[15] that converts the 10 µm laser lines into 16 µm laser lines. The CRISLA patented process uses a continuous CO laser with strong laser output lines in the 5 µm wavelength region [42].

It is assumed that both SILEX and CRISLA processes feed a mixture of UF_6 in a carrier gas with a high heat capacity ratio to provide sufficient supersonic cooling, e.g., xenon. The gas mixture is adiabatically supercooled into a supersonic jet that is laser irradiated to separate the two isotopes (^{235}U and ^{238}U). As shown in Figure 3.12, non-laser-excited molecules will form heavier dimers ($^{238}UF_6$:Xe) that tend to stay in the jet core (the center dotted line). Using a skimmer, they can be separated from the laser-excited lighter monomers or rim gases ($^{235}UF_6$*), which tend to escape from the jet core to the chamber background.

The dimers caught by the skimmer will dissociate into gaseous monomers as soon as the flow returns to subsonic flow and ambient conditions. Both rim gases and core gases are continuously pumped out separately to product and tail reservoirs. The carrier gas is separated in those reservoirs using standard cold-trapping methods [46].

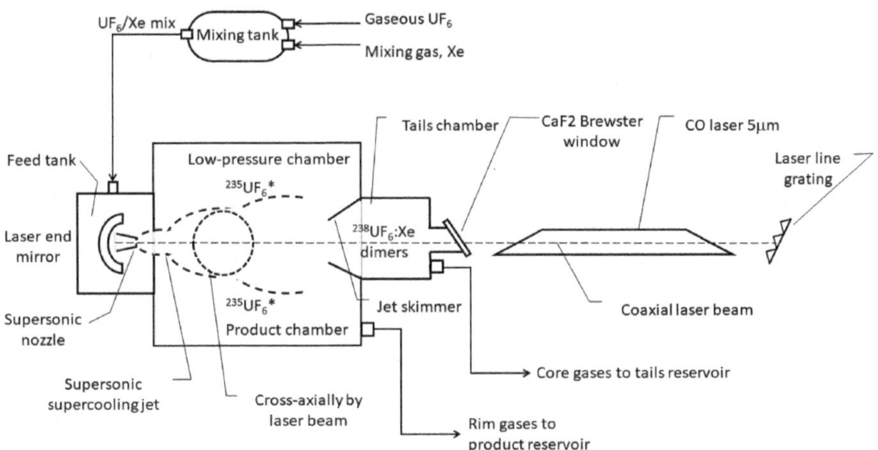

Figure 3.12. Condensation repression separation by laser activation (drawing based on Patent US 10,319,486 B 2).

---

[15] Raman cells are devices used for frequency conversion of fixed or tunable laser sources and the suppression of amplified spontaneous emission. They are based on stimulated Raman-scattering, i.e., the inelastic scattering of photons where there is an exchange of energy, and the scattered photons are shifted in frequency.

Uranium enrichment poses a greater risk of proliferation; hence, these facilities must be particularly safeguarded. Some efforts in this direction have been, for example, the creation of the British-Dutch-German uranium consortium URENCO, governed by the Almelo treaty, to make uranium suitable for use in nuclear power stations in Germany, and the Netherlands; the URENCO and ORANO joint venture in the field of uranium enrichment technology in the US [47]; and EURODIF, a subsidiary of ORANO, that has been operating a facility with four more owners, Italy, Spain, Belgium, and Iran, who have no access at all to the technology [48].

Fuel fabrication

The majority of current commercial nuclear reactors use uranium fuel, natural or enriched, in the form of UO_2 (uranium oxide) as seen in Table 3.1. Natural UO_2 is produced from the purified UO_3 obtained in the conversion process, and enriched $^{235}UO_2$, is produced by reducing the $^{235}UF_6$ from enrichment.

Fuel fabrication is the last stage in the process of fuel preparation and its operations are aimed at achieving accurately shaped ceramic UO_2 pellets and assembling a completed fuel assembly.

The first step is receiving the UO_3 and the $^{235}UF_6$ required for manufacturing. According to regulations [49], the UO_3 arrives in drums type IP-1 for low-specific activity material, and the $^{235}UF_6$, in steel cylinders 30B and an outer protective overpack UX-30 or DN-30 to protect the

Table 3.1. Fuel type according to reactor type.

Reactor type	Fuel type	Countries
Boiling water reactor (BWR)	Enriched UO_2	USA, Japan, Sweden
Pressurized heavy water reactor (PHWR)	Natural UO_2	Canada*, India
Advanced gas-cooled reactor (AGR)	Natural U (metal), enriched UO_2	UK
Light water graphite reactor (LWGR)	Enriched UO_2	Russia
Fast neutron reactor (FBR)	PuO_2 and UO_2	Russia
High-temperature gas-cooled reactor (HTGR)	Enriched UO_2	China
Pressurized water reactor (PWR)	Enriched UO_2	USA, France, Japan, Russia, China, South Korea

* The largest natural uranium fuel producer.

Figure 3.13. $^{235}UF_6$ cylinders inside UX-30 overpacks (Photo from IAEA image bank under Creative Commons license CC BY 2.0).

container from overheating and impacts during transportation. These are illustrated in Figure 3.13. Their design, manufacturing, use, and servicing operations must be consistent with ANSI and ISO standards [50, 51]. The authorization and transport of the packages should also conform to the regulations for the transport of radioactive and fissile materials.

The industrial operations for the manufacturing of natural UO_2 fuel from purified $^{238}UO_3$ are shown in Figure 3.14. The operations to produce low-enriched (3–5% ^{235}U) UO_2 fuel from $^{235}UF_6$ are illustrated in Figure 3.15.

Both processes have in common the production of pure UO_2 powder; the achievement of high-density ceramic UO_2 pellets by granulation and sintering; the loading of pellets into fuel rods, and their bundle assembly into a final structure [52].

The natural uranium fuel fabrication process begins with the reduction of the purified UO_3 in a rotary kiln within a hydrogen-rich atmosphere. This process can be performed using the UO_3 directly or adding water to the UO_3 to form a hydrate.

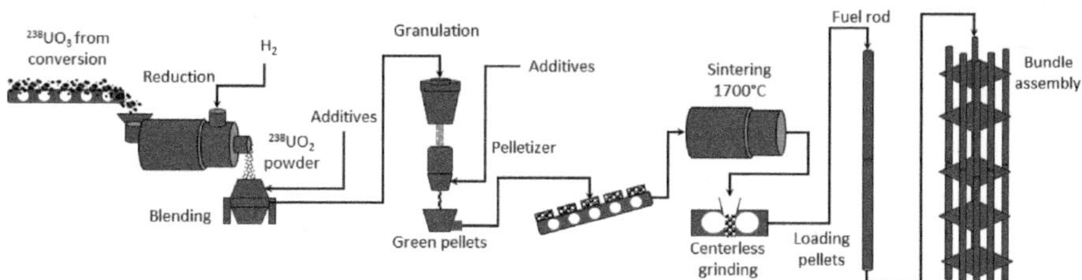

Figure 3.14. Key steps of natural uranium fuel fabrication process.

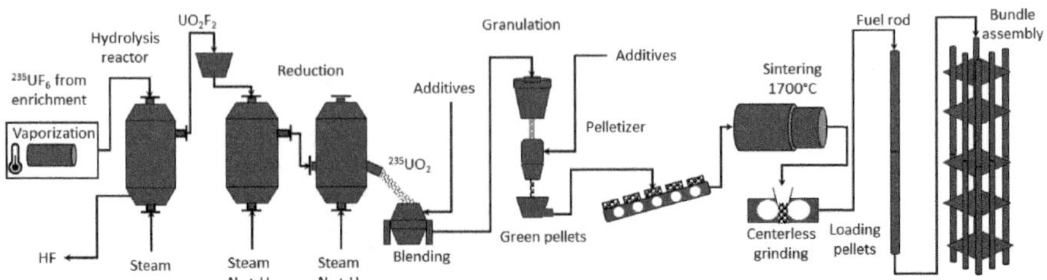

Figure 3.15. Key steps of low-enriched uranium fuel fabrication process.

Low-enriched uranium fuel fabrication process starts with the vaporization of the $^{235}UF_6$ cylinders received from the enrichment facility. Vaporization is performed in steam cabinets where cylinders 30B are heated at about 57°C to release the $^{235}UF_6$ gas, which then is sent to a hydrolysis reactor.

In the hydrolysis reactor, the hexafluoride gas reacts with a steam flow in a fluidized bed of UO_2F_2 particles. These particles overflow to a product hopper and then are transferred to a second reactor, where the bed is fluidized using steam and cracked ammonia ($N_2 + H_2$) to reduce the UO_2F_2 to UO_2. The reduction is repeated in a third reactor to ensure complete conversion to UO_2. The HF resulting from the reaction is recovered using a series of metal filters.

In both processes, the UO_2 powder is prepared for granulation by blending it with die lubricants and small amounts of additives to meet the physical and chemical requirements for granulation. The granulation process ensures a more uniform particle size. Blending and granulation should be performed by limiting the total volume of powder in the blender for nuclear criticality safety [53].

The UO_2 granules are next pressed into compacted green pellets[16] 10–13 mm in length and 8–13.5 mm in diameter. Green pellets are placed on molybdenum sintering boats and fed into a reduced atmosphere furnace for sintering at 1700°C and ~ 4 h to achieve approximately 95% of the theoretical density of UO_2 [54]. Green pellets shrink during sintering, therefore, after cooling, the sintered pellet surface areas are ground, typically using centerless grinders with two wheels, to provide each pellet with the exact size and a slight end taper.[17] Finally, the pellets are washed with water or blown with air to remove any scraps and are thoroughly inspected for surface defects, density, impurities, and dimension accuracy.

UO_2 fuel pellets are moved after inspection to the loading station, where they are inserted into thin zirconium alloy cladding tubes called

---

[16] Green pellets are defined as wet pellets, i.e., non-sintered.

[17] The end taper allows the pellets to expand and contract through drastic temperature changes inside the reactor.

fuel rods. Each fuel rod is sealed at both ends with press-fitted zirconium alloy plugs and automatically welded before the cladding tube is filled with helium. The top-end plug is provided with a spring to keep the pellets in place during handling and shipping. In an inspection area and using visual, radiographic, and ultrasonic methods, the fuel rods are checked and tested for perfect straightness, right dimensions, correct weight, and weld perfection and integrity, among other specifications. Neutron interrogation[18] is also used to assess the enrichment. The fuel rods are finally grouped into fuel assemblies after checking.

Fuel assembly designs are different since they depend on the reactor type. For example, WWER fuel assemblies are hexagonally shaped while PWR fuel assemblies use a square lattice typically of 17 × 17 fuel rods per assembly, and the PHWR assemblies are concentric rings around a central axis. PWR and CANDU assemblies are shown in Figure 3.16.

In addition to the fuel rods, fuel assemblies include empty tubes for control rods or in core instrumentation, nozzles, and tie plates at the top and bottom, and a series of spacer grids to maintain a consistent distance between the fuel rods and the position where the control rods are inserted. BWR fuel assemblies also contain water rods [55, 56].

Fuel assemblies operate in hard conditions where high temperatures, chemical corrosion, radiation, and other physical stresses such as static loads, fluid impacts, vibration, etc., can affect the integrity of their components. Hence the significance of fuel designs with greater fretting wear resistance, improved thermal performance, and better debris filtering efficiency [57].

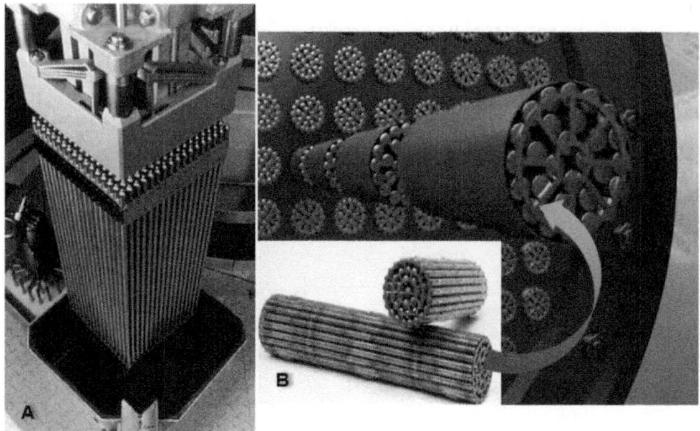

Figure 3.16. (A) PWR nuclear fuel assembly (public photo courtesy of French Alternative Energies and Atomic Energy Commission), (B) CANDU fuel assembly (public photo courtesy of Atomic Energy of Canada Limited).

---

[18] Neutron interrogation is a rapid way of measuring quantities of fissile material.

Fuel burnup during power generation

The fuel burnup, also known as fuel utilization, is a measure of how much uranium is burned in the reactor [58]. In other words, it is the amount of energy extracted per mass of initial fuel loaded, expressed as gigawatt-days per ton of heavy metal loaded (GWd/t) [59].

With current reactors, to maintain the reactor performance, one-third of the fuel is required to be replaced with fresh fuel every 12 or 18 months. The isotopic composition of the spent fuel to be removed will depend on the burnup. US commercial reactors have reached an average discharge burnup of around 50 GWd/t routinely, whereas a maximum of 62 GWd/t has been established for the limit of 5% uranium enrichment [60].

It was observed that cladding corrosion started to increase at higher burnups, and some traditional zirconium alloys had to be replaced by proprietary alloys [61]. Changes in fuel pellet microstructure, higher fission gas release, and rod internal pressures have also been detected, raising safety concerns [62]. Hence, while a further increase to 75 or 80 GWd/t could have some benefits, to receive approval, it requires a technical justification that the fuel remains safe during operation and after removal from the reactor.

Spent fuel interim storage

When removed from the reactor, the spent fuel contains a list of harmful elements, including actinides—uranium, plutonium, neptunium, americium, and curium; and fission products that make it thermally hot and highly radioactive. Therefore, spent fuel assemblies must be unloaded into a storage pool, at the reactor site, similar to the one shown in Figure 3.17, and retained there for many years to cool down.

Figure 3.17. Spent fuel pool at the San Onofre Nuclear Generating Station near San Clemente, California (Photo courtesy of NRC).

A spent fuel storage pool is typically 12 m deep, made with reinforced concrete walls 1.2–2.4 m thick, and with a stainless-steel liner with studs embedded in the concrete. Storage racks, baskets, and containers are used to keep the fuel assemblies at the bottom, with approximately 8 m of water above them [63]. Water is continuously supplied to the pool through water pumps. Extra water, if necessary, may be supplied by pumps connected to an emergency generator. The water level in the pool is continually monitored, as are the temperature and radiation levels inside and around it.

Water pools were planned for short-term storage (5–10 yr) at the beginning. However, countries such as the US, Canada, Germany, Sweden, and Finland, with an open nuclear fuel cycle policy due to concerns of proliferation, public acceptance, and/or economics, have held much of the spent fuel in pools for several decades now. In the US, around 70% of the commercial spent fuel inventory is still stored in wet pools [64].

Since 1986, older spent fuel from pools has been steadily transferred to independent dry spent fuel storage installations (ISFSI) at operating or decommissioned nuclear plant sites, under the responsibility of the plant owner [65]. Currently, there are dry stores in 37 states at 68 sites under general licenses and 17 sites with specific licenses [66]. Both wet and dry storage systems are acceptable. They are designed to shield from radiation, manage heat, and prevent criticality for a period of 50 years [71]. However, cask systems require much less attention and are cheaper than pools.

For example, Figure 3.18 shows the dry storage casks at the Diablo Canyon nuclear plant site. These casks use passive cooling by natural air convection through vents at the bottom. The fuel assemblies are held in sealed helium-filled canisters, inside 6 m tall and 3 m diameter,

Figure 3.18. Dry cask spent nuclear fuel interim storage at Diablo Canyon nuclear plant (photo courtesy of NRC).

Figure 3.19. San Onofre Nuclear Generation Station below-grade spent fuel dry cask storage (photo courtesy of SONGS Decommissioning).

Figure 3.20. Spent fuel horizontal dry storage (photo courtesy of Orano TN International under Creative Commons license).

concrete-filled, steel storage casks, which are also bolted to a 2.3 m thick steel-reinforced concrete pad per seismic requirements [67].

Usually, a transfer cask and a vertical cask transporter are used to load, unload, and transport the spent fuel canister into the dry storage cask. Vertical metal casks can be installed above grade, as in Diablo Cayon, and below grade. For example, Figure 3.19 shows the ventilated vertical underground casks at San Onofre Nuclear Generating Station ISFSI.

Two types of dry casks can be used above ground, the vertical cylindrical casks that were mentioned before, and a series of horizontal rectangular casks as those in Figure 3.20.

In horizontal interim storage designs, the helium-filled, steel-shielded canister holding the spent fuel is placed horizontally inside the concrete module or vault. These storages require additional transfer systems, which may include a transfer cask, a hydraulic ram, and a trailer similar to those in Figure 3.20. The exterior walls and roof of the horizontal vault are 1 m thick reinforced concrete and structural steel, while the interior walls are 60 cm thick reinforced concrete.

In both vertical and horizontal storage alternatives, the fuel is packaged in a form that is compatible with the requirements for transport and/or disposal [68]. Both systems are designed to resist earthquakes, projectiles, tornadoes, floods, temperature extremes, and other scenarios [69]. The inspection of canister surfaces and welds inside the spent fuel dry cask storage systems is possible thanks to innovative crawling robotic technologies [70].

Depending on country policies, used fuel is transferred to central interim storage facilities instead of being kept at nuclear plant sites. For example, in Germany, the Gorleben transport cask storage facility is used for the interim dry storage of spent fuel; in Sweden, the centralized interim storage facility (CLAB) keeps the spent fuel from nuclear plants in various pools about 50 meters underground in granite bedrock; and the outstanding central interim storage facility in Onkalo, Finland, will be the first deep geological repository for high-level nuclear waste that will hopefully begin operating in a year or less.

In the US, a facility, independent from decommissioning and operating nuclear power plants, which is specifically designed to store safely and securely used nuclear fuel until the material can be transported to a national permanent repository is called a consolidated interim storage facility (CISF). The first CISF for spent fuel dry casks is already licensed for construction and operation in Andrews, Texas [71]. A second CISF project in Lea County, New Mexico, is currently under the licensing process with the NRC.

Spent fuel reprocessing

The longer spent fuel is stored, the easier is to handle because of radioactive decay and reduced heat. However, the storage of spent fuel in pools and dry casks is, and should be, just a temporary solution. Presently, it is widely agreed that the only acceptable way to permanently isolate spent fuel and other high-level radioactive nuclear waste from the environment is their disposal in a deep geological repository [72].

Geological disposal could be better supported by a reprocessing process designed to recover the plutonium and uranium from the spent fuel. A fuel cycle with reprocessing implies treating the spent fuel as an asset, not waste. So far, only about one-third of the spent fuel worldwide has been reprocessed [73]. Examples of existing reprocessing capacities are La Hague in France, Rokkasho in Japan, and Tarapur in India.

A choice taken by many countries is to consider the spent fuel as waste to be sent to final disposal without reprocessing. Others have chosen to wait [74]. However, regardless of the progress reached by Finland and Sweden toward a deep final disposal facility, to date, no geological repository has been completed. It is estimated that this option still must go through interim storage for a long period ($\approx$ 100 yr) to allow enough time to define future steps [75].

Before 1977, when reprocessing was banned in the US, three commercial reprocessing plants were in operation. The first, in Idaho Falls, operated successfully from 1966–72; the second, at Morris, Illinois, needed extensive modifications and General Electric (GE) withdrew the operating license in 1974; the third, at Barnwell, South Carolina, was canceled in 1977 after federal support was cut [76]. Hereafter, the efforts have focused on improving the fuel efficiency during reactor operation and the spent fuel interim storage.

According to the Spent Fuel and Radioactive Waste Information System (SRIS) [77], in 2021, only 22.36% of the total tons of heavy metal (tHM) of spent fuel discharged have been reprocessed, while 48.48% still did not have a determined route for management and are waiting in wet or dry interim storages. However, despite the security, proliferation, and safety concerns that have risen in the past, reprocessing is not assumed to be the shortest path to nuclear weapons anymore.

Used fuel from light water reactors contains approximately 96% of U (most of it ^{238}U and < 1.0% ^{235}U); 3% of stable fission products; 1% of Pu produced in the reactor; and < 1% of other fission products and minor actinides [Cs (cesium), Sr (strontium), I (iodine), Tc (technetium), Am (americium), Cm (curium) and Np (neptunium)] [78]. A sizable fraction (> 10%) of spent fuel from these reactors could be reprocessed to extract plutonium, which can be used to make mixed PuO_2 + UO_2 (MOX) fuel for conventional reactors that can be loaded with 1/3 of MOX fuel assemblies.

Plutonium can also be recycled for fast breeder reactors which produce more plutonium than the uranium and plutonium they consume. Russia is leading commercial fast reactor development with the sodium-cooled fast neutron reactor BN-600, which has been supplying electricity to the grid since the early 1980s, and the new BN-800, designed to recover energy from plutonium, which was connected to the grid in 2016 and fully loaded with uranium-plutonium mixed oxide (MOX) fuel in 2022 [79, 80].

Besides, the current development of advanced nuclear reactors is prompting the recycling of spent fuel to reduce the amount of high-level radioactive waste for storage and disposal [81]. Created high-level waste, mostly containing fission products and actinide oxides, is immobilized into very stable borosilicate glass matrixes lowering the risks of release to the environment during storage and disposal. For example, La Hague, which has safely processed spent nuclear fuel from various nuclear power plants around the world in the last 40 years, has the largest vitrification facility in the world [82].

A typical reprocessing procedure like the one known as PUREX is illustrated in Figure 3.21. The first step in reprocessing is cutting up the fuel rods into small pieces to separate the assembly cladding and dissolve the fuel pellets in hot concentrated nitric acid (HNO_3). The cladding material resistant to nitric acid, along with insoluble residues and fuel assembly hardware, remains in the dissolution device. The off-gas containing radioactive iodine, krypton, and xenon volatilized from the dissolution process is trapped for treatment and disposal.

Following the dissolution, a solvent extraction operation is used to separate uranium and plutonium into an organic phase using a mix of tributyl phosphate (TBP) or n-dodecane in kerosene. The fission products (^{90}Sr, ^{137}Cs, ^{99}Tc, ^{126}Sn, ^{129}I, ^{135}Cs, ^{93}Zr) and transuranic minor actinides (^{241}Am, ^{243}Am, ^{240}Pu, ^{239}Pu, ^{237}Np, ^{246}Cm) remain in the aqueous nitric phase and are rerouted for waste treatment. The waste solution is evaporated to reduce its storage volume, calcined, and vitrified into a glass matrix [83].

Uranium and plutonium in the organic phase are taken back to solvent extraction using diluted nitric acid and a reductant, e.g., hydroxylamine, to reduce Pu^{4+} to Pu^{3+} [84]. Separation occurs because of different affinities of uranium and reduced plutonium with the aqueous and organic phases. The aqueous stream of plutonium is concentrated, converted into a solid state by oxalate precipitation (CaC_2O_4), then reduced, and calcined to a

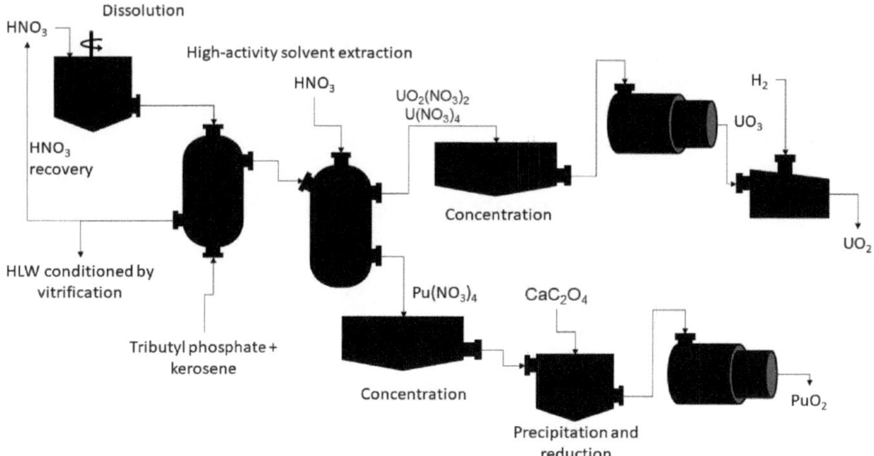

Figure 3.21. Typical reprocessing operations for the recovery of uranium and plutonium.

very pure PuO_2. The aqueous stream of uranium is concentrated, thermally denitrated by calcination, and reduced to UO_2. At least three extractions are usually needed to improve the purity of both uranium and plutonium. In each cycle, fission products are removed and sent to waste treatment.

PUREX is a well-known hydrometallurgical process that was used for several decades in the production of plutonium for military purposes. For over 40 years, PUREX has also been continuously developed and improved by France, Russia, and India for commercial reprocessing without any diversion of special nuclear materials. It demonstrates less cost when compared to the expenses for directly spent fuel disposal; a reduction in volumes and toxicity of the waste for disposal; and an increase in long-term safety for the streams of vitrified HLW [85].

Partitioning and transmutation are methods to potentially impact the long-term disposal of nuclear waste. An advanced fuel cycle with a partitioning of minor actinides and the subsequent transmutation through fissioning would be the most effective technique to reduce the actinide inventory and consequently its radiotoxicity and long-term hazard [86].

Partitioning technologies are being investigated to further remove Am, Cm, Np, Tc, and minor actinides/fission products from the waste streams of reprocessing to further reduce waste radiotoxicity and heat [87]. Research studies include the US Department of Energy project UREX+, where U and then Tc are recovered initially, then Cs and Sr, and the remaining are treated to recover Pu with other transuranic elements; the GANEX process of the French Atomic Energy Commission, which includes a uranium recovery and a simultaneous group separation of actinides; and the US research study TALSPEAK, aimed to separate trivalent lanthanides from trivalent actinides [88].

Radioactive waste transmutation of minor actinides into short-life radionuclides is also being investigated [89, 90]. The goal is to transform a large fraction of spent fuel transuranic actinides into stable or short-lived (< 30 years) radionuclides using a fast neutron reactor or an accelerator-driven device.[19]

Deconversion and re-enrichment

Deconversion is the process of producing chemically stable uranium oxides (U_3O_8 and UO_2) and fluoride products from the depleted uranium hexafluoride (DUF_6) tails generated during uranium enrichment. Depleted uranium oxides could find use in the manufacturing of counterweights, shielding, and military items, or can be disposed of as low-level radioactive waste at existing commercial sites [91]. If long-term storage is however

---

[19] An accelerator-driven system combines a proton accelerator with a specially designed subcritical reactor.

intended, DUF_6 may be deconverted into uranium tetrafluoride (UF_4), which is less volatile and more stable.

In the US, there are around 800,000 t of tails stored in large lines of 14 t steel cylinders 48G, at open-air backyards, in the Paducah and Portsmouth sites, of which only 11% has been treated (less than 100,000 t since 2010) not only because of the COVID-19 pandemic impacts but various inherent technical or mechanical flaws that have forced a frequent shutdown of the deconversion plants [92]. See an illustration of storage cylinders in Figure 3.22. The estimate of DUF_6 tails stored in the world is approximately two million tons.

Cylinders 48G for uranium hexafluoride tail storage are designed according to Title 49 of the Code of Federal Regulations. Its Paragraph 420 [49] states that DUF_6 cylinders shall withstand a hydraulic test at an internal pressure of at least 1.4 MPa without leakage; loss or dispersal of the uranium hexafluoride; and rupture of the containment system. It is also required that the DUF_6 be in solid form and that its volume at 20°C not exceed 62% of the certified volumetric capacity of the cylinder.

The studies performed on these storage cylinders in the late 1990s, after various years of storage, demonstrated just a slight internal corrosion rate of 0.0508 mm/y, apart from the peeling paint, rusting, and weathering from the outdoor exposure seen in Figure 3.22 [93]. A management program is usually organized to ensure the integrity of the cylinders, including periodic visual inspections, cylinder wall thickness measuring, vacuuming debris from the skirt of the cylinders, repainting the ends of skirted cylinders and cylinder bodies, restacking and respacing the cylinders, and improving the storage conditions of the backyards [94]. However, leakages become more likely with the aging of cylinders, and the risk and accidents during storage and handling might increase with time.

Figure 3.22. Stored containers of DUF_6. Peeling paint and rusting are consequences of outdoor exposure (Photo courtesy of the US Department of Energy).

Tricastin, in France, has been deconverting depleted uranium tails since 1984. The more stable depleted triuranium octoxide (U_3O_8) obtained by defluorination can be fed back to fuel fabrication or stored as a future resource for fast breeder reactors [95]. In the UK, at Capenhurst, the Tails Management Facility (TMF) from Urenco ChemPlants, commissioned in 2019, already started to deconvert the stored depleted uranium tails for their long-term storage until final disposal [96]. In the US, the deconversion facilities at Portsmouth and Paducah sites were commissioned in 2010 and 2011, respectively. Their operations include the modification and recycling of the cylinders 48G used for storing DUF_6. The Zelenogorsk defluorination plants in Russia were supplied by Orano, France. The first started production in 2009 and, in 2019, a second plant doubled its capacity. No other countries with enrichment facilities do deconversion [97].

A simplified deconversion process is shown in Figure 3.23. In this example, similar to that described in US Patent 5752158A [98], the UF_6 is first vaporized in an autoclave with steam, at between 80–100°C. Uranyl fluoride (UO_2F_2) is produced in a fluidized bed unit, from the mix of the gaseous depleted uranium fluoride, steam, and hydrogen at ≈ 700°C. The hydrogen fluoride (HF) that is released, passes out the top of the conversion unit and goes into a scrubbing system to be next converted into 70% hydrofluoric acid for industrial use. The UO_2F_2 is next agglomerated and densified in a fluidized bed made of a uranium oxide material. There, it is defluorinated and reduced by countercurrent steam to produce depleted uranium oxides. The resulting depleted uranium is cooled, compacted,

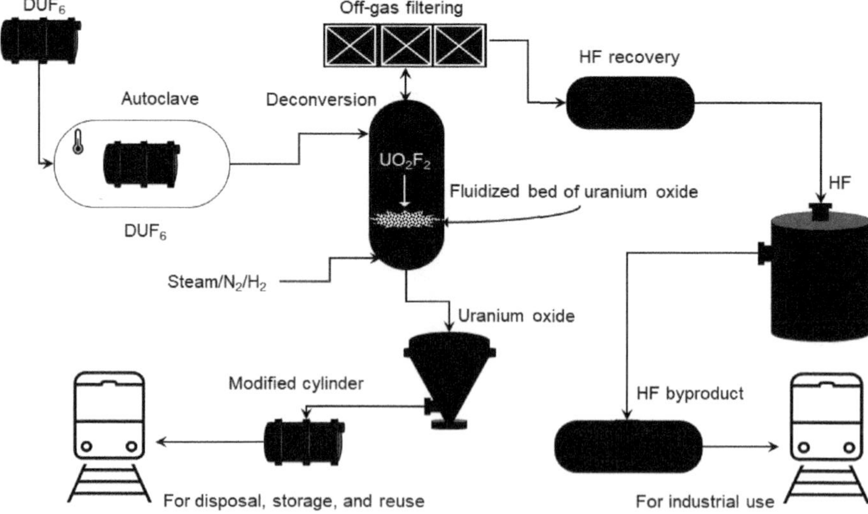

Figure 3.23. Operations of a general DUF_6 deconversion process.

and loaded through hoopers into modified 48G storage containers for eventual disposal or reuse [99].

The modified container is designed to take advantage of the empty containers that are a byproduct of deconversion. These containers are then used for the long-term storage of DU_3O_8 (depleted uranium oxide) [100]. For this purpose, after the vaporization of DUF_6 in the autoclave, the cylinder is cooled down, disconnected, and moved to a station to neutralize its interior with potassium hydroxide (KOH) and to determine its state and residual contamination by inspection. If the cylinder shows no damage or corrosion, it can be modified for the storage of DU_3O_8. To modify a cylinder, a hole is cut at the valve location to make it easier to load uranium oxides, and a patch is then welded over that opening [101].

The deconversion of DUF_6 has been performed in Tricastin, France for many years. First, the DUF_6 is turned into gas using autoclaves. It is next injected into a kiln and heated to 300°C, where it reacts with steam to create HF and UO_2F_2. The HF gas is filtered, condensed, and scrubbed to obtain an aqueous 70% HF, which is transferred to a storage tank for delivery to clients. The solid UO_2F_2 moves to the other part of the kiln on a screw conveyor and reacts with superheated water vapor and additional hydrogen injected by countercurrent to yield depleted triuranium octoxide (DU_3O_8). Finally, the DU_3O_8 is compacted and conditioned before being stored in specially designed 12 t containers DV 70 [102].

As shown in Figure 3.24, DUF_6 might also be deconverted into UF_4, a more stable form than UF_6, and with a much higher temperature of volatilization.

The process in Figure 3.24 uses H_2 and heat to produce DUF_4. It is planned by the US NNSA (National Nuclear Security Administration) to support the production of high-purity DU metal for the military [24],

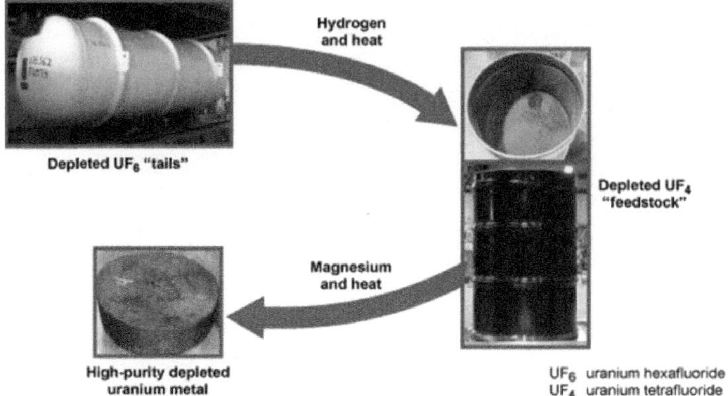

Figure 3.24. Conversion of DUF6 tail into metal (Photo courtesy of the Department of Energy).

[103]. A similar plant, designed to deconvert DUF_6 tails into DUO_2 and DUF_4 anhydrous, was licensed by the NRC in 2012 in Lea, New Mexico. This was planned to transform DUF_4 into high-purity SiF_4, and/or BF_3, along with anhydrous HF as a byproduct [104]. However, no construction activities have occurred.

Another way to treat the accumulated DUF_6 tails is by a re-enrichment process to recover and reuse them. The GLE Enrichment Facility in Paducah is using SILEX technology to re-enrich DUF_6 tails to natural-grade $^{238}UF_6$. The NRC licensed this project and it seems to be a great prospect [105]. It involves processing over 200,000 t of tails in the next 30–40 yr [106]. The first full-scale laser system demonstration was completed in Australia in 2022 and is now to be tested in the Loop Facility at Wilmington, North Carolina, in the US [107].

References

[1] World Nuclear Association. 2022. Uranium Enrichment. World Nuclear Association, October 2022. [Online]. Available: https://world-nuclear.org/information-library/nuclear-fuel-cycle/conversion-enrichment-and-fabrication/uranium-enrichment.aspx. [Accessed 13 April 2022].

[2] US NRC. 2020. Deconversion of Depleted Uranium. US Nuclear Regulatory Commission, 2 December 2020. [Online]. Available: https://www.nrc.gov/materials/fuel-cycle-fac/ur-deconversion.html. [Accessed 13 April 2022].

[3] World Nuclear Association. 2023. World Uranium Mining Production. World Nuclear Association, August 2023. [Online]. Available: https://world-nuclear.org/information-library/nuclear-fuel-cycle/mining-of-uranium/world-uranium-mining-production.aspx [Accessed 12 October 2023].

[4] Cameco Corporation. 2022. Cameco's Uranium 101. Cameco Corporation, 2022. [Online]. Available: https://www.cameco.com/uranium_101/uranium-overview/finding-uranium/. [Accessed 18 April 2022].

[5] Uranium Producers of America. 2014. Conventional Mining and Milling of Uranium Ore. Uranium Producers of America, 2014. [Online]. Available: http://www.theupa.org/uranium_technology/conventional_mining/. [Accessed 21 April 2022].

[6] New Mexico Bureau of Geology & Mineral Resources. 2022. Uranium & Nuclear Energy. New Mexico Institute of Mining & Technology, 22 February 2022. [Online]. Available: https://geoinfo.nmt.edu/resources/uranium/home.html. [Accessed 15 April 2022].

[7] Cameco Corporation. 2022. Cameco's U 101 Mining and Milling: Mining. Cameco Corporation, 2022. [Online]. Available: https://www.cameco.com/uranium_101/mining-milling/mining/. [Accessed 24 April 2022].

[8] International Atomic Energy Agency. 2020. Occupational Radiation Protection in the Uranium Mining and Processing Industry, Safety Report Series No. 100, Vienna: IAEA.

[9] International Atomic Energy Agency. 2000. IAEA-TECDOC-1174, Methods of exploitation of different types of uranium deposits, Vienna: IAEA.

[10] Newman, G., L. Newman, D. Chapman and T. Harbicht. 2011. Artificial Ground Freezing: An Environmental Best Practice at Cameco's Uranium Mining Operations in Northern Saskatchewan, Canada. In 11th International Mine Water Association Congress – Mine Water – Managing the Challenges, Aachen.

[11] Donev, J. et al. 2021. Energy Education - Uranium mining. University of Calgary, 18 October 2021. [Online]. Available: https://energyeducation.ca/encyclopedia/Uranium_mining [Accessed 24 April 2022].

[12] Yun, X., K. McNamara and G. Murdock. 2011. Geotechnical challenges and strategies at McArthur river operation. Procedia Engineering 26(7): 1603–1613, December 2011.

[13] Cameco Corporation. 2022. Mining Methods. Cameco Corporation, 2022. [Online]. Available: https://www.cameco.com/businesses/mining-methods [Accessed 22 April 2022].

[14] International Atomic Energy Agency. 2016. Nuclear Energy Series No. NF-T-1.4. *In Situ* Leach Uranium Mining: An Overview of Operations. IAEA, Vienna.

[15] Garside, M. 2023. Uranium production worldwide from 2014 to 2022, by mining method. Statista, 29 August 2023. [Online]. Available: https://www.statista.com/statistics/202358/world-uranium-mine-production-by-method/#statisticContainer. [Accessed 13 October 2023].

[16] International Atomic Energy Agency. 1993. Uranium Extraction Technology, IAEA Technical Report No. 359, Vienna: IAEA.

[17] Zhu, Z. and C. Y. Cheng. 2011. A review of uranium solvent extraction: its present status and future trends. In ALTA 2011 Uranium Conference: 7th Uranium Event, Perth.

[18] International Atomic Energy Agency. 2004. IAEA-TECDOC-1419, Treatment of liquid effluent from uranium mines and mills. Report of a co-ordinated research project 1996–2000. IAEA, Vienna.

[19] Rajan, K., G. Ramadevi, G. Nitai, J. Chakravartty and T. Sreenivas. 2016. Studies on conversion of crude sodium diuranate to high purity uranium oxide by chemical route. In Proceedings of the Seventh DAE-BRNS Biennial Symposium on Emerging Trends in Separation Science and Technology, Guwahati.

[20] Murty, B. N. and H. Ravindra. 2013. Analysis of sodium diuranate for organic flocculent content. Journal of the Indian Chemical Society 90(11): 2035–2038.

[21] ASTM International. 2021. C788-03(2021) Standard Specification for Nuclear-Grade Uranyl Nitrate Solution or Crystals, West Conshohocken: ASTM International.

[22] International Atomic Energy Agency. 2018. Regulations for the Safe Transport of Radioactive Material 2018 Edition. Specific Safety Requirements No. SSR-6 (Rev. 1). IAEA, Vienna.

[23] US NRC. 2010. Module 3.0 of the Fuel Cycle Processes Directed Self-Study Course: Conversion, Chattanooga: USNRC Technical Training Center.

[24] World Nuclear Association. 2022. Conversion and Deconversion. World Nuclear Association, January 2022. [Online]. Available: https://world-nuclear.org/information-library/nuclear-fuel-cycle/conversion-enrichment-and-fabrication/conversion-and-deconversion.aspx. [Accessed 3 May 2022].

[25] United States Government Accountability Office (GAO). 2020. Report to Congressional Committees GAO-21-28, Uranium Management, Actions to Mitigate Risks to Domestic Supply Chain Could Be Better Planned and Coordinated. GAO, Washington.

[26] McGinnis, B. 2013. ORNL/TM-2013/72 An Overview of Process Monitoring Related to the Production of Uranium Ore Concentrate. Oak Ridge National Laboratory, Oak Ridge.

[27] Raffo-Caiado, A. C., J. M. Begovich, M. A. Saraiva Marzo, L. C. Palhares et al. 2009. Model of a Generic ORNL/TM-2008/195 Natural Uranium Conversion Plant—Suggested Measures to Strengthen International Safeguards. Oak Ridge National Laboratory, Oak Ridge.

[28] Figueroa, J. and M. A. Williamson. 2008. ANL/CSE-13/25 Uranium Dioxide Conversion. Argonne National Laboratory, Argonne.

[29] Cameco Corporation. 2015. 2017 License Renewal Application for the Port Hope Conversion Facility, Saskatoon: CAMECO.

[30] Amamoto, I., T. Terai, H. Oobayashi and R. Fujita. 2002. Behaviour of impurities in recycled uranium at uranium conversion process. Journal of Nuclear Science and Technology 39(Sup. 3): 769–771.

[31] Lawrence Livermore National Laboratory. 1992. UCRL-ID-120925, Evaluation of a Dry Process for Conversion of U-AVLIS Product to UF6 Milestone U361. US DOE, Livermore.

[32] Federal Government. 2022. Title 10 Chapter I Part 71 Packaging and Transportation of Radioactive Material. National Archives Code of Federal Regulations, 1 June 2022. [Online]. Available: https://www.ecfr.gov/current/title-10/chapter-I/part-71?toc=1. [Accessed 7 June 2022].

[33] International Atomic Energy Agency. 2022. Water-cooled reactors. IAEA, n.d. [Online]. Available: https://www.iaea.org/topics/water-cooled-reactors. [Accessed 7 May 2022].

[34] Energy.gov. 2022. Paducah GDP Shutdown and Deactivation. Portsmouth/Paducah Project Office. U.S. Department of Energy, n.d. [Online]. Available: https://www. energy.gov/pppo/paducah-gdp-shutdown-and-deactivation#:~:text=In%20May%20 2013%2C%20the%20United,the%20leased%20facilities%20to%20DOE. [Accessed 30 September 2022].

[35] US NRC. 2020. Uranium Enrichment. US Nuclear Regulatory Commission, 2 December 2020. [Online]. Available: https://www.nrc.gov/materials/fuel-cycle-fac/ur-enrichment.html. [Accessed 15 May 2022].

[36] Chemistry LibreTexts. 2020. 2.9: Graham's Laws of Diffusion and Effusion. The LibreTexts libraries, 1 September 2020. [Online]. Available: https://chem. libretexts.org/Bookshelves/Physical_and_Theoretical_Chemistry_Textbook_Maps/ Map%3A_Physical_Chemistry_for_the_Biosciences_(Chang)/02%3A_Properties_of_ Gases/2.9%3A_Graham's_Laws_of_Diffusion_and_Effusion. [Accessed 16 May 2022].

[37] World Nuclear News. 2023. Enrichment operations start at US HALEU plant. World Nuclear Association, 12 October 2023. [Online]. Available: https://www.world-nuclear-news.org/Articles/Enrichment-operations-start-at-US-HALEU-plant. [Accessed 14 October 2023].

[38] Centrus Energy. 2022. American Centrifuge. Centrus Energy Corp., 2022. [Online]. Available: https://www.centrusenergy.com/what-we-do/national-security/american-centrifuge/. [Accessed 16 May 2022].

[39] SILEX. 2022. Uranium Enrichment Technology. Silex Systems Limited, 24 February 2022. [Online]. Available: https://www.silex.com.au/silex-technology/silex-uranium-enrichment-technology/. [Accessed 17 May 2022].

[40] Knief, R. A. 2003. Nuclear Fuel Cycles III.E.3 Laser. pp. 655–670. In Encyclopedia of Physical Science and Technology (Third Edition), New York, Academic Press Inc.

[41] Snyder, R. 2016. A proliferation assessment of third generation laser uranium enrichment technology. Science & Global Security 24(2): 68–91.

[42] Eerkens, J. W. 2019. Process and Apparatus for Condensation Repressing Isotope Separation by Laser Activation. United States of America Patent 10,319,486, 11 June 2019.

[43] Goldsworthy, M. 2022. Silex Systems Limited. Investor Presentation. Silex Systems Ltd, 24 February 2022. [Online]. Available: https://www.silex.com.au/silex-technology/ silex-uranium-enrichment-technology/. [Accessed 19 May 2022].

[44] GLE. 2022. The Evolution of Laser Enrichment and Formation of GLE. Global Laser Enrichment, 2022. [Online]. Available: https://www.gle-us.com/history/. [Accessed 20 May 2022].

[45] Silex Systems Limited. 2022. Operational Update. Lucas Heights Science and Technology Centre, Lucas Heights.

[46] Eerkens, J. W. and J. Kim. 2010. Isotope separation by selective laser-assisted repression of condensation in supersonic free jets. American Institute of Chemical Engineers Journal 56(9): 2331–2337.

[47] Urenco. 2017. News. Urenco's commitment to the Treaty of Almelo. Urenco, 14 June 2017. [Online]. Available: https://www.urenco.com/news/global/2017/urencos-commitment-to-the-treaty-of-almelo. [Accessed 5 June 2022].

[48] National Academy of Sciences. 2009. National Research Council, Internationalization of the Nuclear Fuel Cycle: Goals, Strategies, and Challenges, Washington: The National Academies Press.

[49] Federal Government. 2022. Title 49 Subtitle B Chapter I Subchapter C Part 173 Subpart I Class 7 (Radioactive) Materials. National Archives Code of Federal Regulations, 31 May 2022. [Online]. Available: https://www.ecfr.gov/current/title-49/subtitle-B/chapter-I/subchapter-C/part-173/subpart-I?toc=1. [Accessed 7 June 2022].

[50] ANSI. 2019. ANSI N14.1-2019 Uranium Hexafluoride - Packaging for Transport, New York: American National Standards Institute, Inc.

[51] ISO. 2020. ISO 7195:2020 Nuclear energy — Packagings for the transport of uranium hexafluoride (UF6), Geneva: International Organization for Standardization.

[52] World Nuclear Association. 2021. Nuclear Fuel and its Fabrication. World Nuclear Association, October 2021. [Online]. Available: https://world-nuclear.org/information-library/nuclear-fuel-cycle/conversion-enrichment-and-fabrication/fuel-fabrication.aspx. [Accessed 15 June 2022].

[53] US NRC. 2010. Module 5.0 of the Fuel Cycle Processes Directed Self-Study Course: Fuel Fabrication, Chattanooga: USNRC Technical Training Center.

[54] Wood, E. S., J. T. White, B. Jaques, D. Burkes and P. Demkowicz. 2020. Advances in fuel fabrication. pp. 371–418. In Advances in Nuclear Fuel Chemistry, Cambridge, Woodhead Publishing.

[55] Abe, T. and K. Asakura. 2012. Uranium Oxide and MOX Production. pp. 393–422. In Comprehensive Nuclear Materials Vol. 2, Elsevier.

[56] Nuclear Energy Agency. 1998. Committee on the Safety of Nuclear Installations. NEA/CSNI/R(98)20 VVER-Specific Features Regarding Core Degradation Status. Nuclear Energy Agency, Paris.

[57] International Atomic Energy Agency. 2010. Nuclear Energy Series No. NF-T-2.1 Review of Fuel Failures in Water Cooled Reactors STI/PUB/1445. IAEA, Vienna.

[58] US NRC. 2018. Backgrounder on High Burnup Spent Nuclear Fuel. US Nuclear Regulatory Commission, 1 October 2018. [Online]. Available: https://www.nrc.gov/reading-rm/doc-collections/fact-sheets/bg-high-burnup-spent-fuel.html. [Accessed 12 June 2022].

[59] Worrall, A. 2014. Core and fuel technologies in integral pressurized-water reactors. pp. 79–102. In Handbook of Small Modular Nuclear Reactors, Cambridge, Woodhead Publishing.

[60] US NRC. 2020. Accidents Tolerant Fuel Regulatory Activities. Higher Burnup. US Nuclear Regulatory Commission, 13 November 2020. [Online]. Available: https://www.nrc.gov/reactors/atf/burnup.html. [Accessed 10 June 2022].

[61] Geelhood, K. 2019. PNNL-29368, Fuel Performance Considerations and Data Needs for Burnup above 62 GWd/MTU. In-Reactor Performance, Storage, and Transportation of Spent Nuclear Fuel. Pacific Northwest National Laboratory, Richland.

[62] Adamson, R., B. Cox, F. Garzarolli, A. Strasser, P. Rudling and G. Wikmark. 2003. High Burnup Fuel Issues. Advanced Nuclear Technology International, Surahammar.

[63] National Academies of Sciences, Engineering, and Medicine. 2006. Background on Spent Fuel Pool Storage. pp. 38–59. In Safety and Security of Commercial Spent Nuclear Fuel Storage Public Report, Washington, National Academy Press.

[64] Werner, J. D. 2012. U.S. Spent Nuclear Fuel Storage CRS R42513. Congressional Research Service, Washington.

[65] Federal Government. 2022. Title 10 Chapter I Part 72 Licensing requirements for the independent storage of spent nuclear fuel, high-level radioactive waste, and reactor-related greater than Class C waste. National Archives Code of Federal Regulations, 21 June 2022. [Online]. Available: https://www.ecfr.gov/current/title-10/chapter-I/part-72?toc=1. [Accessed 23 June 2022].

[66] US NRC. 2023. Dry Cask Storage. Nuclear Regulatory Commission, 16 June 2023. [Online]. Available: https://www.nrc.gov/waste/spent-fuel-storage/dry-cask-storage.html. [Accessed 18 October 2023].

[67] Diablo Canyon Decommissioning Engagement Panel. 2022. Spent Fuel Management. Diablo Canyon Decommissioning Engagement Panel, 2022. [Online]. Available: https://diablocanyonpanel.org/decom-topics/spent-fuel-management/. [Accessed 24 June 2022].

[68] Hambley, D., A. Laferrere, W. S. Walters, Z. Hodgson, S. Wickham and P. Richardson. 2016. Lessons learned from a review of international approaches to spent fuel management. EPJ Nuclear Sciences & Technologies 2(26).

[69] US NRC. 2021. Backgrounder on Dry Cask Storage of Spent Nuclear Fuel. US Nuclear Regulatory Commission, March 2021. [Online]. Available: https://www.nrc.gov/reading-rm/doc-collections/fact-sheets/dry-cask-storage.html. [Accessed 25 June 2022].

[70] Nuclear Engineering International. 2021. Orano completes inspection of NUHOMS nuclear fuel store. Progressive Media International, 10 February 2021. [Online]. Available: https://www.neimagazine.com/news/newsorano-completes-inspection-of-nuhoms-nuclear-fuel-store-8507363. [Accessed 30 June 2022].

[71] US NRC. 2021. NRC Issues License to Interim Storage Partners for Consolidated Spent Nuclear Fuel Interim Storage Facility in Texas, Washington: NRC News Office of Public Affairs.

[72] Nuclear Threat Initiative (NTI). 2015. Developing Spent Fuel Strategies Project. NTI, Washington.

[73] Czerwin, J. and C. Evans. 2019. Valuing Flexibility and Integrating Risks in Used Nuclear Fuel Management. In Management of Spent Fuel from Nuclear Power Reactors, Learning from the Past, Enabling the Future, Proceedings of an International Conference, Vienna.

[74] International Atomic Energy Agency. 2005. Technical Report No. 425, Country Nuclear Fuel Cycle Profiles, Second Edition, Vienna: IAEA.

[75] International Atomic Energy Agency. 2020. Specific Safety Guide SSG-15 Rev 1 Storage of Spent Nuclear Fuel, Vienna: IAEA.

[76] Andrews, A. 2008. CRS Report for Congress. Nuclear Fuel Reprocessing: U.S. Policy Development. Congressional Research Service, Washington DC.

[77] International Atomic Energy Agency. 2021. Spent Fuel Summary. IAEA-European Union, n.d 2021. [Online]. Available: https://sris.iaea.org/region-overview/spent-fuel-summary. [Accessed 19 October 2023].

[78] World Nuclear Association. 2020. Processing of Used Nuclear Fuel. World Nuclear Association, December 2020. [Online]. Available: https://world-nuclear.org/information-library/nuclear-fuel-cycle/fuel-recycling/processing-of-used-nuclear-fuel.aspx. [Accessed 25 May 2022].

[79] Kütt, M., F. Frieß and M. Englert. 2014. Plutonium disposition in the BN-800 fast reactor: An assessment of plutonium isotopic and breeding. Science & Global Security 22(3): 188–208.

[80] World Nuclear News (WNN). 2022. Beloyarsk BN-800 fast reactor running on MOX. World Nuclear Association, 13 September 2022. [Online]. Available: https://www.

world-nuclear-news.org/Articles/Beloyarsk-BN-800-fast-reactor-running-on-MOX. [Accessed 31 October 2022].

[81] World Nuclear News (WNN). 2022. New US program to investigate recycling of used fuel. World Nuclear Association, 16 March 2022. [Online]. Available: https://www. world-nuclear-news.org/Articles/New-US-programme-to-investigate-used-fuel-recycle [Accessed 29 July 2022].

[82] Wang, P. 2017. La Hague Nuclear Recycling and Reprocessing Plant. Stanford University, Winter 2017. [Online]. Available: http://large.stanford.edu/courses/2017/ ph241/wang2/#:~:text=%28Source%3A%20P.%20Wang%29%20Nuclear%20 recycling%20and%20reprocessing%20allows,volume%20of%20waste%20that%20 is%20must%20be%20disposed. [Accessed 31 October 2022].

[83] Irish, E. R. and W. H. Reas. 1957. The PUREX Process - a Solvent Extraction Reprocessing Method for Irradiated Uranium. U.S. Atomic Energy Commission, Washington.

[84] Paviet-Hartmann, P., C. Riddle, K. Campbell and E. Mausolf. 2013. Overview of Reductants Utilized in Nuclear Fuel Reprocessing/Recycling. Idaho National Laboratory, Idaho Falls.

[85] International Atomic Energy Agency. 2008. IAEA TECDOC 1587, Spent Fuel Reprocessing Options, Vienna: IAEA.

[86] International Atomic Energy Agency. 2004. Technical Reports Series No. 435, Implications of Partitioning and Transmutation in Radioactive Waste Management, STI/DOC/010/435, Vienna: IAEA.

[87] Bourg, S., A. Geist, J.-M. Adnet, C. Rhodes and B. C. Hanson. 2020. Partitioning and transmutation strategy R&D for nuclear-spent fuel: the SACSESS and GENIORS projects. EPJ Nuclear Sciences & Technologies 6(35).

[88] World Nuclear Association. 2020. Processing of Used Nuclear Fuel. World Nuclear Association, December 2020. [Online]. Available: https://world-nuclear.org/ information-library/nuclear-fuel-cycle/fuel-recycling/processing-of-used-nuclear-fuel. aspx. [Accessed 31 October 2022].

[89] Sun, X., W. Luo, H. Lan et al. 2022. Transmutation of long-lived fission products in an advanced nuclear energy system. Scientific Reports 12: 2240.

[90] Richter, B., D. C. Hoffman, S. K. Mtingwa, R. P. Omberg, J. L. Rempe and D. Warin. 2006. Advanced Nuclear Transformation Technology Subcommittee of the Nuclear Energy Research Advisory Committee. Department of Energy, Washington.

[91] US NRC. 2015. Frequently Asked Questions about Depleted Uranium Deconversion Facilities. US Nuclear Regulatory Commission, 23 March 2015. [Online]. Available: https://www.nrc.gov/materials/fuel-cycle-fac/ur-deconversion/faq-depleted-ur-decon.html. [Accessed 4 November 2022].

[92] US Department of Energy Office of Inspector General. 2022. Audit Report on Depleted Uranium Hexafluoride Conversion Operations DOE-OIG-23-04. U.S. Department of Energy, Washington.

[93] Glenn, M. L. 1974. Uranium Hexafluoride Tails Storage Cylinders, Report Number: KY-657. Union Carbide Corporation Nuclear Division, Paducah.

[94] Argonne National Laboratory. 2022. Depleted UF6. U.S. DOE Office of Environmental Management [Online]. Available: https://web.evs.anl.gov/uranium/mgmtuses/ cylsurv/index.cfm. [Accessed November 2022].

[95] Orano. 2022. International expert in uranium processing. Orano Group, [Online]. Available: https://www.orano.group/en/nuclear-expertise/from-exploration-to-recycling/international-expert-in-uranium-processing. [Accessed 13 November 2022].

[96] Urenco ChemPlants. 2022. Urenco ChemPlants, a wholly owned Urenco subsidiary, owns and will operate our Tails Management Facility in the UK. Urenco, 2022. [Online]. Available: https://www.urenco.com/global-operations/urenco-chemplants. [Accessed 13 November 2022].

[97] World Nuclear Association. 2021. Russia's Nuclear Fuel Cycle. World Nuclear Association, December 2021. [Online]. Available: https://www.world-nuclear.org/information-library/country-profiles/countries-o-s/russia-nuclear-fuel-cycle.aspx. [Accessed 13 November 2022].

[98] Urza, I. J. and R. Wash. 1989. Conversion of Uranium Hexafluoride to Uranium Dioxide. United States of America Patent 4,830,841, 16 May 1989.

[99] Urza, J. and K. Barlow. 2005. Stabilization of Depleted Uranium Hexafluoride Tails Material for Disposition. In Waste Management 2005 Conference, Tucson, 2005.

[100] Energy.gov. 2022. DUF6 Conversion Project Portsmouth/Paducah Project Office. US Department of Energy [Online]. Available: https://www.energy.gov/pppo/pppo-services/pppo-cleanup-projects-portsmouth-paducah-duf6/duf6-conversion-project. [Accessed 8 November 2022].

[101] O'Connor, D. G., A. B. Poole and J. H. Shelton. 2000. ORNL/CF-OO/36, Assessment of Reusing 14-t, Thin-Wall, Depleted UF6 Cylinders as LLW Disposal Containers. Oak Ridge National Laboratory, Oak Ridge.

[102] Duperret, B., A. Maillard and B. Le Motais. 2005. Management of Depleted UF6 300,000 t DUF6 Defluorinated Lesson Learned. In Waste Management 2005 Conference, Tucson.

[103] GAO. 2020. Nuclear Weapons: NNSA Plans to Modernize Critical Depleted Uranium Capabilities and Improve Program Management GAO-21-16. U.S. Government Accountability Office, Washington.

[104] US NRC. 2020. IIFP Fluorine Extraction and Depleted Uranium Deconversion Plant Licensing. Nuclear Regulatory Commission, 2 December 2020. [Online]. Available: https://www.nrc.gov/materials/fuel-cycle-fac/inisfacility.html. [Accessed 19 October 2023].

[105] US NRC. 2020. Global Laser Enrichment Facility Licensing. Nuclear Regulatory Commission, 2 December 2020. [Online]. Available: https://www.nrc.gov/materials/fuel-cycle-fac/laser.html. [Accessed 19 October 2023].

[106] Goldsworthy, M. 2022. AGM Presentation: Silex Systems Limited 2022 Annual General Meeting. www.silex.com.au, Lucas Heights.

[107] World Nuclear News WNN. 2022. Testing complete for laser enrichment module. World Nuclear Association, 5 September 2022. [Online]. Available: https://www.world-nuclear-news.org/Articles/Testing-complete-for-laser-enrichment-module. [Accessed 19 November 2022].

CHAPTER 4

Safety in Nuclear Fuel Cycle Facilities

Safety in nuclear fuel facilities encompasses several aspects, including radiation safety[1] (radiation protection and nuclear safety), criticality safety,[2] industrial safety, and nuclear security.[3] Consequently, a comprehensive array of requirements, procedures, devices, and equipment is necessary to avert unacceptable risks to individuals and the environment. These risks arise from the radioactive materials that are treated, handled, stored, and prepared for disposal; the involvement of chemical substances and gases that may be toxic, corrosive, combustible, or reactive; the possibility for a criticality accident;[4] and potential non-proliferation and security threats.

National authorities have set specific requirements for all facilities involved in the nuclear fuel cycle and spent fuel storage. These requirements aim to protect workers, the public, and the environment. Such requirements have also been agreed upon by international organizations such as the International Atomic Energy Agency (IAEA), and the Nuclear Energy Agency (NEA) of the Organization of Economic Cooperation and Development (OECD), along with the Food and Agriculture Organization (FAO), the International Labor Organization (ILO), the International Maritime Organization (IMO), the Pan American Health Organization (PAHO), the United Nations Environment Program (UNEP) and the World Health Organization (WHO) [1, 2, 3].

---

[1] The protection of people and the environment against radiation risks, and the safety of facilities and activities that give rise to radiation risks.

[2] The prevention and protection from an uncontrolled nuclear fission chain reaction or a criticality accident.

[3] The prevention and detection of criminal or intentional unauthorized acts involving nuclear material and other radioactive material.

[4] Accidental uncontrolled nuclear fission chain reaction as a result of the accumulation of a critical mass. (See footnote 15 in Chapter 1.)

The above-mentioned facilities include: open pit, underground, and *in-situ* leach mines for mining and processing[5] of uranium ores; yellowcake (U_3O_8) fabrication centers; facilities for dry and wet conversion of uranium into UF_6; centrifugal (or new generation laser) ^{235}U enrichment facilities; and the facilities for the fabrication of nuclear fuel itself into precisely formed ceramic pellets, fuel rods, and fuel assemblies. It also includes locations for the interim storage of fissile and fertile materials, and the management of radioactive waste—waste conditioning, effluent treatment, and waste storage. Where applicable, facilities include those for reprocessing spent fuel to recover uranium and other transuranic elements.

Authorization process

From facility siting to decommissioning and release from regulatory control, independent authorizations are required for the various stages a nuclear fuel facility goes through. For instance, site approval is required to prove its suitability after considering every possible adverse event—of natural origin or human-induced—on the environment, and neighboring populations. Design approval and construction authorization confirm their conformity with national laws and regulations while operating, and shutdown licenses prove the safety of the planned activities. Likewise, authorizations are required for the transport of radioactive materials— in approved containers—as well as the management of radioactive waste and effluents.

Radiation safety outline

Protection of workers and the public from radiation is the fundamental objective of the safety of nuclear fuel cycle facilities. As a naturally occurring radioactive element, uranium is an alpha emitter in secular equilibrium[6] with its progeny. However, during the mining and processing of uranium ores, this equilibrium may be disrupted and potential exposure to γ radiation from its progeny could be possible. Dermal exposure can occur when the skin comes into direct contact with uranium dust. Additionally, internal exposure can occur if uranium dust or radon decay products are inhaled or ingested. Significant quantities of highly enriched material (^{235}U) can also pose a γ radiation hazard [3, 4, 5].

---

[5] Processing includes both physical and chemical procedures that alter the ore after it is removed from its deposit in nature.

[6] When the parent nucleus has an extremely long half-life like ^{238}U, an equilibrium is established where progeny nuclides are decaying at the same rate as they are being produced, so their quantity remains constant.

The most important decay products from the point of view of radiation protection are ^{234}U, ^{230}Th, ^{226}Ra, ^{210}Pb (lead), and ^{210}Po (polonium). Gaseous ^{222}Rn and its progeny—^{218}Po, ^{214}Pb, ^{214}Bi (bismuth), and ^{214}Po—are also critical because of their ability to enter into the working atmosphere and cause worker exposure. ^{214}Bi is the major contributor to gamma exposure [6].

If inhalation and/or ingestion occurs, alpha and beta particles from uranium and its progeny are deposited inside the body, directly irradiating the tissues, and increasing the probability of damage. Internal exposure to high levels of radium, e.g., may lead to bone cancer, while inhaled radon and its short-lived progeny accumulate in the respiratory tract damaging the mucous lining, which may lead to lung cancer [7, 8]. In addition, nearly all uranium compounds are highly toxic and can affect the kidneys, liver, and lungs [9].

The organization and management of the radiation protection program for the facility should be approached for the specific process of the fuel cycle at hand and should be commensurate with the likelihood and magnitude of radiation exposures. Some general requirements will be addressed next as a summary of the scope of the most important radiation protection measures.

The buildup of long-lived radionuclide dust is usually expected in ore milling and the production of yellowcake, while potential inhalation and ingestion of radon progeny are more likely in underground mining. A key requirement to prevent the accumulation of dust and gases is the use of ventilation, including high-efficiency air filtration systems in work areas, and before discharge to the environment. Ventilation is critical in underground mines, uranium mills, uranium product packaging and storage areas, locations where radioisotope vaporization could occur due to high temperatures, and so on. Some areas will need negative air pressure to keep the source of contamination contained, i.e., a differential pressure that would only allow drawing air from cleaner places. Enclosed operating booths with filtered air supplies can also be useful, e.g., to house the staff who carry out remote control operations in underground mines.

Along with proper ventilation, continuous checking through air sampling and fixed radiation monitoring instruments in working areas, booths, and some enclosed spaces should be arranged to verify that the concentration of radon progeny and airborne dust are at safe levels, consistent with established regulations for the air quality in the workplace and ambient air.

Because of discomfort and possible interference with efficiency, the routine use of personal respiratory protection is not recommended. It is instead required in specific tasks, e.g., dusting maintenance jobs; in areas with high concentrations of radon progeny; non-enclosed operations with possible decay product buildup; emergencies; emergency recovering actions; or when handling dry uranium concentrate; etc. [6].

Another general requirement is the classification of areas as controlled or supervised, in correspondence with the expected risks.

Controlled areas can be airborne radioactive areas, radiation areas, high or very high radiation areas, and/or places where radioactive material is used or stored. These areas require compliance with detailed procedures, the use of personal dosimeters, and the supply of fixed and portable equipment for the measurement and control of radiation doses and contamination. Controlled areas must be delimited by physical barriers, e.g., gates, fences, walls, corridors, etc., or occupy separate buildings. They should be labeled at the entrance with the radiation warning symbol shown in Figure 4.1.

To prevent the spread of contamination, controlled areas are equipped with access control zones. These zones are furnished with wardrobes for changing clothes and are equipped with contamination monitoring instruments to detect radioactive contamination on personnel clothes and protective equipment (PPE). Typically, these access control zones also include washing facilities, such as handwashing sinks and showers, to differentiate clean and potentially contaminated areas from the entrance [10]. The dimensions and quantity of these zones will depend on the processes, the levels of radiation, and the contamination involved.

Some examples of controlled areas are digging locations inside the mine; uranium milling and crushing areas; the chemical plant that extracts the uranium; spaces where the uranium product is precipitated, filtrated,

Figure 4.1. Basic ionizing radiation symbol as specified by the international standard ISO R 361:1975, reviewed and confirmed in 2020.

dried, weighed, and packaged; product storage areas; spent fuel storage areas; and so forth.

Supervised areas are those places, typically in the surroundings of the controlled areas, which are commonly occupied by personnel and require monitoring to detect any deviation from the normally expected exposure, i.e., an elevated background, as a result of any control failure. Areas such as nearby workshops, workers' eating areas, laboratories, and the like can be examples of supervised areas. Also, the depleted uranium open-air backyards.

The basic radiation protection approach of time–distance–shielding applies to the process design and day-to-day production of uranium. It is a crucial approach to minimize the dose workers could receive in any area.

Aside from task rotation to restrict the time a person spends in high-dose areas and/or in direct contact with the ore, or the product, it is recommended that workers wear an individual alarm dosimeter, so they know when to leave the place. Today, there are small-sized electronic dosimeters of continuous readout, which data can be quickly downloaded to a computer, e.g., in the control zone.

All processes, e.g., yellowcake fabrication, conversion, precipitation, solvent extraction, ion exchange, etc., should be designed to keep workers in low-dose areas most of the time. To avoid unnecessary exposure, spaces like offices, workshops, and laboratories, as well as eating, resting, and smoking areas should be located at a certain distance from process capacities (which could refer to equipment, apparatus, or tools), stockpiles, and tailings piles.

Remote-controlled equipment, e.g., scooptrams to remove the extracted ore and clean the waste rock on floors in underground mining; chutes to transport the ore and slurries through a pipe network; and *in-situ* leach mining and processing, are examples of operations that can be performed at a distance and without worker's exposure.

Shielding can easily be integrated into the engineering design and is normally built into equipment and other work systems. In underground mining, for instance, both the heavy mining equipment and the shotcrete[7] used in ground support, provide shielding against gamma radiation. Shielding is also part of room structures, material handling tools, safety enclosures, pipes, tanks, vessels, cylinders, containers, and so forth. Mobile shielding barriers, and removable shields such as lead sheets, and bricks

---

[7] Shotcrete is a type of concrete that is sprayed onto a surface at a high velocity, forming a strong bond with whatever it is applied to.

may be used too. In the spent fuel interim storage pools, water works as shielding against neutrons and γ radiation.

Glove boxes provide shielding while ensuring containment to handle radioactive materials in laboratories, test facilities, and manufacturing sites. They are sealed booths with forced filtered ventilation, a viewing window, and gloves built into the side or front to work inside the enclosure. Where higher doses are involved, reinforced concrete hot cells with robotic manipulators and tools are required.

To keep doses from external and internal exposures as low as reasonably achievable, it is a requirement to set operational limits for normal operation and predicted occurrences in the specific nuclear fuel cycle facility or facilities. Operational limits are usually below dose limits and they apply to all operations involved in any particular fuel cycle process, including the management of radioactive waste and releases. Predicted incidents are previously assessed and emergency arrangements are put together to mitigate their consequences in case they occur [11].

The dose limit is the value of the effective dose to individuals that is not to be exceeded in any planned exposure situation. The limit internationally recommended for occupational exposure—workers' exposure—is 20 mSv per year averaged over five consecutive years and 50 mSv in any single year. The dose limit for members of the public— persons who may occasionally come into contact with effluents or waste released from the facility, including other employees in the vicinity of the controlled areas, is 1 mSv in a year [12].

In the US, limits for public and occupational exposure are established by Title 10, Part 20, of the Code of Federal Regulations [13]. For workers, the annual total effective dose equivalent for the whole body is 50 mSv. For the public, the limit is an additional 1 mSv in a year above the natural radiation background. Some historical doses can illustrate the performance of these limits. For example, the data from 17 US uranium mills around 1975 (before the initial period of moderate growth of uranium mining) show an average annual dose of 3.8 mSv, while the data from all mines in Canada in the period 2010–2014 shows an average annual dose of 1.1 mSv. Both are well below the annual limit for radiation workers [14].

Some data on worldwide workers' exposure in the nuclear fuel cycle sector for the period 2000-2014 is listed below. The estimated average annual effective dose in that period was about 0.6 mSv, which is below the limit but, in most cases, higher than the doses to occupational workers in other sectors of radioactive materials applications [15].

Nuclear fuel cycle operation	Average annual effective dose (mSv)[8]
In situ leaching extraction of uranium	3.9
Underground uranium mining	1.8
Open-pit uranium mining	1.5
Open-pit uranium processing	1.4
Underground uranium processing	0.5
Other techniques	0.9
Uranium enrichment	0.1
Uranium fuel fabrication	0.9
Safe management of spent fuel	0.2–11
Spent fuel reprocessing	0.2
Radioactive waste management	0.2–3
Transport of nuclear material	0.3–0.7

From the list above, it is noteworthy that workers from *in-situ* leaching (ISL) mining operations show higher annual dose values. This was attributed, as mentioned in the report, to an overestimation of the annual dose because of no correction for dosimeter background exposure [15]. However, it should be pointed out that radon gas release from the lixiviant could be an issue without adequate ventilation in the occupied areas. Radium can also build up in pipework and vessels, contributing to external exposure to gamma radiation [16].

To prevent accidents and mitigate their consequences if they occur, a defense-in-depth approach is required. This concept, which will be discussed further in Chapter 6, is based on the application of multiple independent and redundant barriers and levels of protection to all operations, including those with chemical hazards. These measures are designed to compensate for potential human and technical failures. If a failure were to occur at any one of these levels, it would be detected and either compensated for or corrected through the application of measures at the other levels [17].

Minimizing the volume and activity of radioactive waste is a critical condition for protecting the public and the environment. This implies that waste should only be disposed of once its generation has been minimized

---

[8] This unit represents the average dose absorbed in a given tissue or organ adjusted by radiation type and relative organ sensitivity. The SI unit for an effective dose is joule per kilogram (J/kg), called the sievert (Sv). 1 mSv = 0.001 Sv.

Figure 4.2. Exhausted JEB (jet bored) open pit turned into a tailings impoundment in Saskatchewan, Canada (Photo courtesy of Canadian Nuclear Safety Commission).

and all potential recycling materials are reused. The efficient management of radioactive waste and effluents is a radiation protection requirement for all nuclear fuel facilities. This requirement applies from waste generation to its final disposal, including pretreatment, treatment, conditioning, storage, and transport.

An example of a dam for the disposal of uranium tailings is illustrated in Figure 4.2. This impoundment, constructed using an old Canadian open pit uranium mine, is specifically designed to reduce the potential water table contamination and prevent radon release and dispersion of radioactive dust.

Measures to reduce the gas releases into the environment to the minimum achievable or eliminate them include off-gas treatment capacities such as electrostatic precipitators, scrubbers, chemical traps, and HEPA filters. These capacities, along with ventilation filtration and cleaning systems, as well as air monitoring techniques and procedures are crucial to ensure safety in all processes of the fuel cycle.

Conditioning of sludges is another way to isolate and secure radioactive waste from the environment and human intrusion. Several streams of waste can be conditioned in a matrix of bitumen or concrete for disposal as low or intermediate-level radioactive waste. Borosilicate glass matrixes are used for liquids and sludges of higher activity. Radioactive waste storage requires shielding as well as airborne and liquid effluent radioactivity monitoring. In some cases, cooling, and criticality control, along with pool water chemistry are also needed.

Safety analysis

Achieving the best design and construction, as well as the safe operation of nuclear fuel cycle installations, requires the analysis and assessment of specific potential scenarios that allow for the reciprocal action and superposition of chemical, radiological, and nuclear hazards. In general, these scenarios respond to the following initiating events [11]:

- Loss of services (electrical power, compressed air, coolant, etc.)
- Loss of criticality controls (a drop of fuel, loss of geometry, loss of neutron poison, flooding,[9] change of phase, etc.)
- Processing errors (lack or excess of reagent or coolant, slow or late addition, rupture of pressure retaining vessel or pipe, extreme temperature, etc.)
- Facility and equipment failures (loss of confinement, leakage, filter or column blockage or bypass, item false actuation, etc.)
- Handling errors (safety interlock or lifting failure, load drop, etc.)
- Other internal events (fire, explosion, flooding, malfunction, criticality event, etc.)
- External events (earthquakes, tornados, hurricanes, storms, aircraft crashes, etc.)
- Human errors (wrong specifications, operator or maintenance error or omission, etc.)

The likelihood and severity of those events will affect the planning and implementation of safety. Therefore, if any postulated event happens to occur, the facility could be secured by the action of features and systems that are still available or have been brought into operation in response to the event. These include emergency preparedness and response procedures.

Outline of chemical hazards

The use of aggressive chemicals in nuclear fuel cycle facilities under harsh environmental conditions—heat, pressure, dust, and complex mixtures— may cause exothermic reactions leading to overheating, and even fire or explosions, which must be managed to avert an uncontrolled release of relatively large quantities of radioactive materials. Consequently, the removal of heat from radioactive decay and chemical reactions is one of the most important safety functions.

---

[9] An abnormal presence of a quantity of fluid, either in the form of accumulation, flow, or spray.

Furthermore, preventing the accumulation of flammable or explosive materials such as hydrogen and other gases is a safety measure that should be accomplished with proper local ventilation and cooling. Extreme care is advised to prevent the accumulation of radiolytic hydrogen in the high-activity waste tanks from reprocessing.

Local ventilation is also required to remove the finely divided oxidizable dust or pyrophoric materials that could be generated in operations like evaporation, nitric acid dissolution, or organic-matter reactions, in addition to maintaining the concentration of fluorine and other hazardous gases below prescribed limits.

Aside from the hazards related to the gas and dust accumulation, other adverse scenarios such as vibrations, the reach of extreme temperatures and pressures from reactions, or the possible release of fluids from failed systems or mechanisms in process vessels, are to be considered to prevent the ignitability and reactivity of the chemicals involved [18].

During the storage and transport of hazardous chemical materials, it is important to meet the requirements for their correct containment, segregation, and cooling. Hazardous materials should not be stored together with radioactive materials, nor present where radioactive materials are handled. Flammable materials must be restricted in process areas, switch rooms, and control rooms.

Fluorine is also a strong oxidizing agent, characterized by its toxicity, corrosiveness, and reactivity. It is a requirement then, that materials used for vessels, processing equipment, confinement, and containment are resistant to highly corrosive chemicals such as F_2 and HF, as well as wear-resistant to facilitate cleaning and the removal of radioactive contamination.

Nuclear criticality safety outline

In a fuel cycle facility, providing shielding to protect against an accidental uncontrolled fission chain reaction, i.e., a critical excursion, may be not always feasible. Hence, the focus should be on anticipating events that can lead to such an excursion and acting in advance to prevent them.

A critical excursion implies a sudden release of energy in the form of heat and ionizing radiation that could be lethal to nearby personnel. Such excursion can also damage process equipment and cause the release of radioactive material into the environment. Therefore, a nuclear fuel facility requires a series of preventive measures to maintain the process conditions. It also requires mitigation measures to reduce the overall risk [19].

Criticality safety must be integrated into the design unless the amount of fissile material is very low, or the isotopic composition meets the exemption criteria approved by the regulatory authority. Criticality safety should be taken into account in the selection of the chemical products to be used, and in the administrative controls to be applied. The most important parameters to consider in criticality safety are [20]:

- The mass,[10] volume, and enrichment of fissile materials
- The size and shape of pipes, vessels, units, and processing equipment
- The concentration of fissile materials in solutions
- The conditions that allow precipitation, and
- The degree of neutron moderation and reflection

Consistent with the double contingency principle, the design of the process should include sufficient safety factors to ensure that a criticality accident can only occur if there are at least two unlikely, independent, and simultaneous changes in process conditions [21].

Therefore, the mass, volume, and enrichment of fissile materials in the process should be within the limits established by administrative procedures after careful assessment. It is a requirement to use instruments to measure and control the concentration of the solutions to prevent any inadvertent change that may lead to precipitation, colloid formation, and increases in concentration in the solutions.

Following the same line of reasoning, the geometries of the units and equipment in the processes (their size and shape) should not be greater than the subcritical limit on the capacity of the vessel or container. Generally, each vessel or container is labeled with the volume or mass limit.

It is important to be aware that a criticality accident can be initiated by placing a shielding nearby, mixing the fissionable materials with water or acids, or even changing the location of the vessel or container. The analysis of scenarios leading to an accidental criticality, together with the existence of reliable administrative and engineered controls to guarantee that such events are highly unlikely, are crucial requirements for criticality safety [22].

The analysis is accomplished assuming failures related to important parameters, e.g., loss of geometry; loss of neutron poison; a drop of fuel during handling; surplus of reflection or moderation; multiple batching and flooding; loss of confinement; transitory configurations during transfers; neutron interaction between facility units; loss of electrical power and compressed air; and so forth.

If it can be demonstrated that the inherent features of the process, equipment, or material in a given accident sequence constrain the reactivity of fissile material within subcritical limits, such accidental

---

[10] It is important to bear in mind that the safe mass of uranium varies with enrichment.

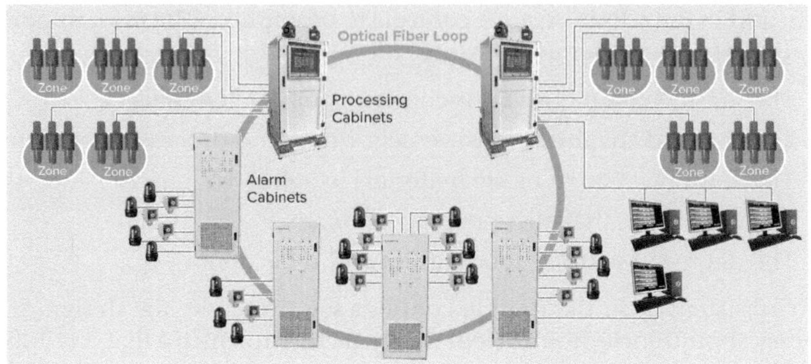

Figure 4.3. Criticality accident alarm system CAAS-3S from Mirion Technologies (Product Image courtesy of Mirion Technologies (Canberra), Inc.).

criticality can be considered not credible [23]. For example, the probability of a criticality accident in plants processing only low-enrichment uranium (LEU) is inherently extremely low.

It is well known that, in the case of a nuclear criticality accident, there is not expected to be any indication that an accident occurred or is about to occur. Therefore, the availability of an accidental criticality detection and alarm system is the most important mitigative measure [24].

An accidental criticality detection and alarm system comparable to the one shown in Figure 4.3 consists of several gamma and neutron radiation detectors, with audible and visible alarms, covering multiple areas where a significant quantity of fissile materials is present. The alarms will alert personnel to promptly evacuate the area to prevent likely radiation exposure. In the emergency plans, the identification of evacuation routes and regrouping areas, after an alarm is triggered, should be an important part of the mitigation measures.

The interface between security and safety in the nuclear fuel cycle

Safety, security, and non-proliferation are integral components of the use of nuclear energy, aligning with its three basic principles. These principles prescribe that nuclear energy must be beneficial, responsible, sustainable, and aimed solely at one essential common objective—the effective utilization of nuclear energy while safeguarding people, society, and the environment. As shown in Figure 4.4, these components are interconnected and should be incorporated into both the facility design and the management system [25].

For instance, preventing and detecting the diversion of nuclear material is not only the focus of the safeguards system but also a task

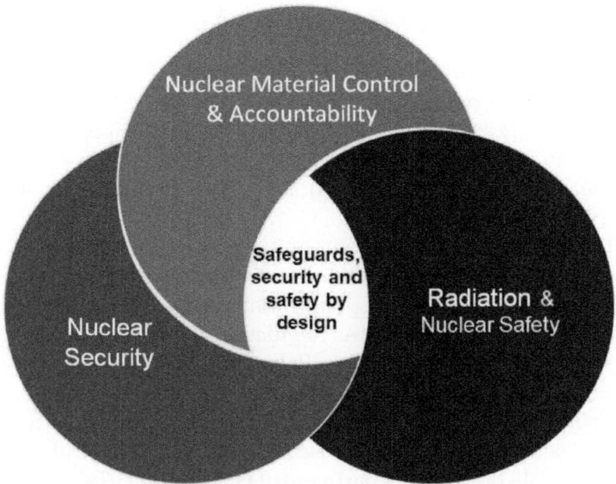

Figure 4.4. Interface between safety, security, and control, and accountability of nuclear materials.

of the security regime. This is a particular concern in enrichment and reprocessing facilities or where nuclear materials are stored. Similarly, many measures, such as limiting access, could serve as a safety measure to prevent worker exposure and a security action to delay or prevent intruders.

The various levels of protection, including prevention measures, early detection, mitigation, and emergency planning are considered in both the safety and security risk assessments made for the design and operation of the facility, commensurate with its potential hazards.

While the main focus of safety is on risks arising from unintended events initiated by, for example, natural occurrences, hardware failures, fires, pipe breakages, or human errors, the primary focus of security is to protect against malicious acts, including theft and sabotage, which may also lead to unacceptable radiological consequences.

The aims may differ, but the means of both safety and security could be very similar and should be designed and implemented in an integrated manner. For instance, the same passive system used in safety to avoid human errors could serve as a security device to make tampering with safety systems more difficult.

The safety analysis of potential failures and their consequences can also be used to identify security-sensitive targets. Personnel screening and access controls increase staff reliability and prevent unauthorized persons from acting. The inventory of radioactive material is a measure of safeguards but also a way to control their safety and security. An efficient management system should ensure a good balance and coordination of all interfaces.

References

[1] International Atomic Energy Agency. 2017. Safety of Nuclear Fuel Cycle Facilities, Specific Safety Requirements No. SSR-4. IAEA, Vienna.

[2] International Atomic Energy Agency. 2020. Storage of Spent Nuclear Fuel, Specific Safety Guide SSG-15 (Rev. 1). IAEA, Vienna.

[3] OECD Nuclear Energy Agency (NEA). 2005. The Safety of the Nuclear Fuel Cycle. OECD, Paris.

[4] US EPA. 2022. Radioactive Waste From Uranium Mining and Milling. US Environmental Protection Agency, July 2022. [Online]. Available: https://www.epa.gov/radtown/radioactive-waste-uranium-mining-and-milling. [Accessed 1 December 2022].

[5] US NRC. 2022. Radium. US Nuclear Regulatory Commission, 21 July 2022. [Online]. Available: https://www.nrc.gov/materials/radium.html. [Accessed 1 December 2022].

[6] International Atomic Energy Agency. 2020. Safety Reports Series No. 100, Occupational Radiation Protection in the Uranium Mining and Processing Industry, Vienna: IAEA.

[7] Agency for Toxic Substances and Disease Registry. 1990. Public Health Statement Radium CAS#: 7440-14-4. U.S. Department of Health and Human Services, Washington.

[8] Agency for Toxic Substances and Disease Registry. 2012. Toxicological Profile for Radon. U.S. Department of Health and Human Services, Washington.

[9] Agency for Toxic Substances and Disease Registry. 2013. Toxicological Profile for Uranium. U.S. Department of Health and Human Services, Washington.

[10] Domenech, H. 2017. Radiation Safety. Management and Programs, Switzerland: Springer International.

[11] International Atomic Energy Agency. 2017. IAEA Safety Standards, Specific Safety Requirements No. SSR-4, Safety of Nuclear Fuel Cycle Facilities, Vienna: IAEA.

[12] International Atomic Energy Agency. 2014. Radiation Protection and Safety of Radiation Sources: International Basic Safety Standards, IAEA Safety Standards Series General Safety Requirements Part 3 No. GSR part 3., Vienna: IAEA.

[13] Federal Government. 2023. Title 10 Chapter I Part 20 Standards for Protection Against Radiation, Washington.

[14] Brown, S. H. 2016. Comparison of Worker and Public Doses from Conventional Uranium Mining and Milling in North America. SHB Inc., Centennial.

[15] United Nations Scientific Committee on the Effects of Atomic Radiation (UNSCEAR). 2022. Sources, Effects, and Risks of Ionizing Radiation, 2020/2021 Report Volume IV Scientific Annex with Appendices Annex D: Evaluation of Occupational Exposure to Ionizing Radiation. United Nations, New York.

[16] International Atomic Energy Agency. 2020. Safety Report Series No. 100. Occupational Radiation Protection in the Uranium Mining and Processing Industry. IAEA, Vienna.

[17] OECD Nuclear Energy Agency (NEA). 2021. Management and Disposal of High-Level Radioactive Waste: Global Progress and Solutions. OECD, Paris.

[18] International Atomic Energy Agency. 2021. Specific Safety Guide No. SSG-64. Protection against Internal Hazards in the Design of Nuclear Power Plants, Vienna: IAEA.

[19] US NRC. 2012. Module 4.0 Nuclear Criticality Safety 0905 (Rev. 3); Nuclear Criticality Safety Controls, Chatanooga: USNRC Technical Training Center.

[20] Costa, D. A., M. E. Cournoyer, J. F. Merhege, V. E. Garcia and A. N. Sandoval. 2017. A primer on criticality safety. Journal of Chemical Health and Safety 24(3): 7–13.

[21] ANSI/ANS. 2014. American National Standard ANSI/ANS-8.1-2014 Nuclear Criticality Safety In Operations With Fissionable Material Outside Reactors, New York: American National Standards Institute (ANSI)/American Nuclear Society (ANS).

[22] Federal Government. 2023. Title 10 Chapter I Part 70, Domestic Licensing of Special Nuclear Material. National Archives Code of Federal Regulations, 3 January 2023. [Online]. Available: https://www.ecfr.gov/current/title-10/chapter-I/part-70. [Accessed 8 January 2023].

[23] Beckner, W. and R. Pierson. 2004. Use of Less Than Optimal Bounding Assumptions in Criticality Safety Analysis at Fuel Cycle Facilities. USNRC, Washington.

[24] International Atomic Energy Agency. 2022. Specific Safety Guide No. SSG-27 (Rev. 1). Criticality Safety in the Handling of Fissile Material, Vienna: IAEA.

[25] International Atomic Energy Agency. 2008. IAEA Nuclear Energy Series No. NE-BP, Nuclear Energy Basic Principles, STI/PUB/1374, Vienna: IAEA.

CHAPTER 5

Overview of Nuclear Reactors

The nuclear reactor is the proper environment for a chain controlled reaction to occur. It is the heart of a nuclear power plant. As seen in the simplified depiction in Figure 5.1, the heat produced by fission reaction in a nuclear reactor is used to convert water into pressurized steam in a steam generator like in any other power plant. This steam then spin a turbine[1] that drives a generator[2] to produce the electricity that is used, e.g., to power cities [1].

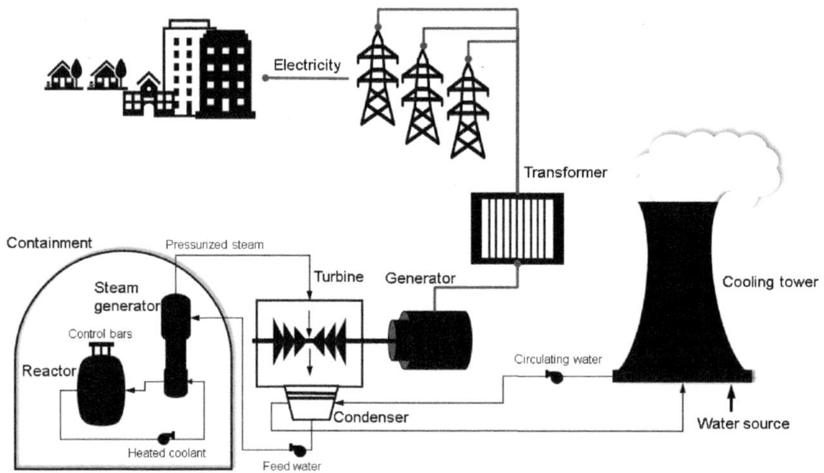

Figure 5.1. Simplified schema of how electricity is produced by a nuclear power plant.

---

[1] A machine that extracts thermal energy from steam and uses it to do mechanical work on a rotating output shaft.

[2] A device that converts the mechanical energy from the turbine into electric power for use in an external circuit.

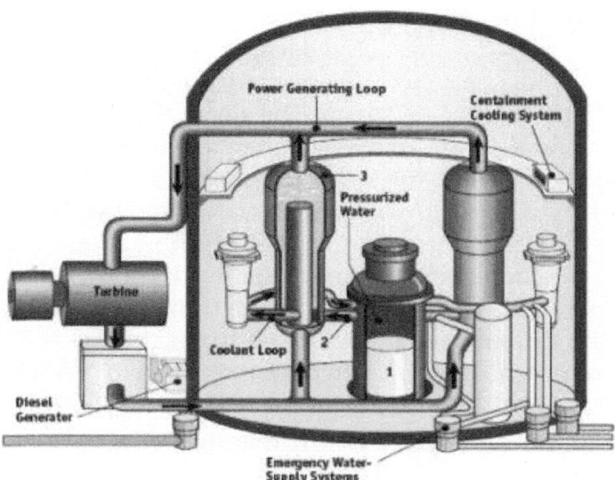

Figure 5.2. Diagram of a pressurized-water nuclear reactor (Image courtesy of the US Nuclear Regulatory Commission).

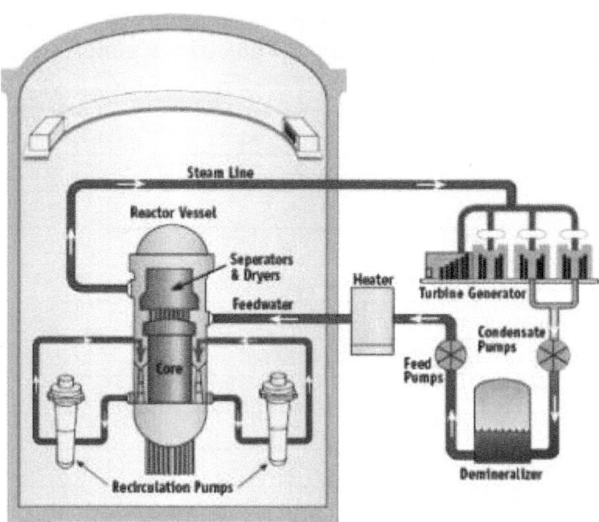

Figure 5.3. Diagram of a boiling-water nuclear reactor (Image courtesy of the US Nuclear Regulatory Commission).

The majority (96%) of the currently commercially operating nuclear power reactors are pressurized water reactors (PWR) and boiling water reactors (BWR), both of which use water as a coolant and moderator. How they work is illustrated in Figures 5.2 and 5.3 [2].

In Figure 5.2, the heat from the core inside the PWR is carried to the steam generator by the pressurized water[3] in the primary loop. In the steam generator, the heat from the primary loop vaporizes the water in a secondary circuit to produce steam. The steam from the generator is then directed to the turbine. The excess steam in the turbine is then condensed, reheated, and sent back to the steam generator.

In Figure 5.3, the low-pressure water cooling the core of the BWR moves upwards through the core absorbing heat and creating a steam-water mixture. This mixture exits the top of the core to enter a separator-dryer that removes the moisture from the steam, which is then directed to the turbine. The excess steam in the turbine is also sent to be condensed, reheated, and pumped back to the reactor vessel.

Some statistics on nuclear reactors

As per recent data, there are more than 400 nuclear reactors in operation in 34 countries, producing a total nuclear energy capacity of around 360 gigawatts-electric (GWe) [3]. The countries with the largest number of nuclear reactors are the US, France, and China, however, France has the greatest share of nuclear power in total electricity generation.

Country[4]	Number of reactors in operation	Share in total electricity generation
USA	92	19.6%
*France	56	69%
China	55	5%
Russia	37	20%
Republic of South Korea	25	28%
India	22	3.2%
Canada	19	14.3%
Japan	17	7.2%
Ukraine	15	55%

---

[3] The water should be pressurized to prevent boiling.
[4] There are over 30 countries with nuclear power plants, but those with less than 10 reactors in operation are not shown in the list.

The number and total capacity (GWe) of operating nuclear reactors by reactor type as of June 2023 are presented below.

Nuclear reactor type	Number*	GWe
Pressurized light-water-moderated and cooled reactors (PWR)	301	287.7
Boiling light-water-cooled and moderated reactors (BWR)	41	43.1
Pressurized heavy-water-moderated and cooled reactors (PHWR)	46	24.1
Light-water-cooled graphite-moderated reactors (LWGR)	11	7.4
Gas-cooled, graphite-moderated reactors (GCR)	8	4.7
Fast breeder reactors (FBR)	2	1.4
High-temperature gas-cooled reactor (HTGCR)	1	0.2
Total	410*	368.6

* Note that the data is dynamic, and the current statistics change over time.

Nuclear reactor types and their components

The basic classification of nuclear reactors is thermal and fast reactors depending on the neutron energy. However, most of the reactors in operation are thermal reactors, and only a couple of them are fast reactors. Thermal reactors are complex systems with many components. These are largely known and classified by the coolant and the moderator they use.

Around 97% of currently operating reactors use light water as the coolant, and water or graphite as moderators. PWRs, BWRs, and VVERs (Russian PWR) are water-water reactors, meaning water-cooled and water-moderated. The Russian LWGR, generally known as RBMK, uses water for cooling and graphite as the moderator. The British AGR reactor is gas-cooled and graphite-moderated. Canadian CANDU reactors use heavy water both for cooling and as the moderator.

The reactor vessel, usually cylindrical, open at the top, and made of metal up to 30 cm thick, is the largest component of a nuclear reactor. It is designed to contain the reactor core, i.e., the bounded region where the fission reaction occurs. The core usually includes the fuel assemblies, the moderator, neutron poisons, control rods, and support structures.

The coolant is the fluid that cools down the core and transfers the heat to produce steam, as explained before. In addition to the primary cooling loop, redundant and diversified emergency core cooling systems are available to limit the damage if a loss-of-coolant accident[5] (LOCA) occurs. The moderator (water, graphite) is used to slow down the neutrons arising

---

[5] Per NRC definition, a potential accident in which a breach in a reactor's pressure boundary causes the coolant water to rush out of the reactor faster than makeup water can be added back in.

from a fission reaction that are of high-velocity or put in other way, the fast neutrons are slowed down to thermal energy neutrons. Commercial reactors are thermal reactors. It means that inside the nuclear reactor, the fast neutrons are slowed down to the thermal energies via a process called neutron moderation.

PWRs, BWRs, LWGRs, and AGRs are fueled using ceramic pellets of low-enriched ($\approx$ 3–5 wt.%) $^{235}UO_2$, while heavy water reactors operate on natural UO_2. Some 40 commercial light-water reactors in France, Germany, Switzerland, Belgium, and Japan are currently using MOX fuel, which is a mixture of PuO_2 and natural, reprocessed, or depleted UO_2 [4].

A scheme of a PWR is shown in Figure 5.4. It is the most common commercial reactor in operation in the world. The UO_2 pellets are loaded within rods of zirconium alloy and into the fuel assemblies in a precise grid pattern, to be arranged vertically in the core, inside the steel pressure vessel. The water, used as a coolant and moderator, is pressurized at about 15 MPa to prevent it from boiling at high operating temperatures. Rods bearing neutron absorbers are inserted into the core or withdrawn from it to adjust the rate of fission. Russian PWR reactors, known as VVER, only differ from PWRs in the hexagonal arrangement of the fuel assemblies.

The BWR is illustrated in Figure 5.5. In these reactors, the pressure inside the vessel is low enough to allow the boiling of the coolant within the core. The steam created mixes with the boiling water and goes through a separator and a dryer, before being directed to the turbine. It also uses fuel assemblies with ceramic pellets of enriched UO_2 encased in rods. Each BWR fuel assembly is enclosed in a zircaloy sheath, or channel box, filled with water to increase the amount of moderator in the central region of the assembly. Control rods are inserted from the bottom of the reactor [5].

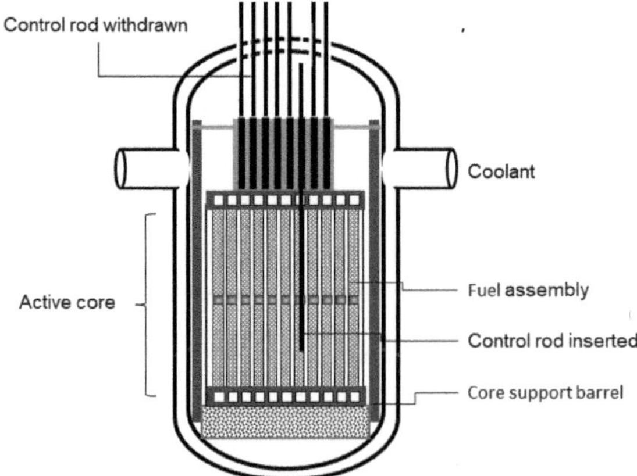

Figure 5.4. A pressurized water reactor (PWR) vessel and components. It uses enriched uranium as fuel and light water as a coolant and moderator.

As sketched in Figure 5.6, the RBMK core is made of graphite with around 1700 vertical channels, each with a fuel assembly. The graphite stack, made of graphite bricks, is inside a cylindrical steel vessel, enclosed at the top and bottom by upper and lower biological shields [6]. The fuel channels and drive channels for control rods are loaded into the graphite stack. Pressurized water—at about 6.9 MPa—is pumped into the graphite channels to remove the heat and create steam. The mix of

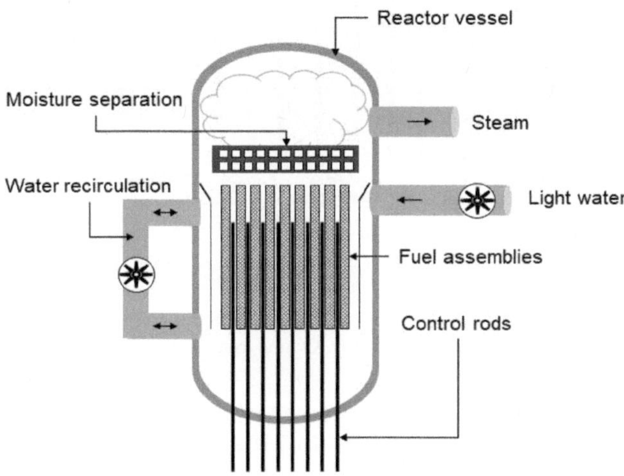

Figure 5.5. A boiling water reactor (BWR) vessel and components. It uses enriched uranium as fuel and light water as a coolant and moderator.

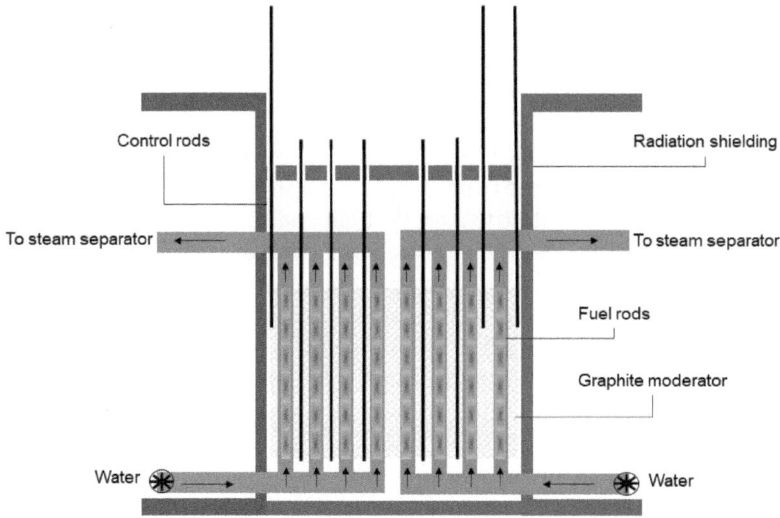

Figure 5.6. Schema of an RBMK reactor. It uses enriched uranium as fuel, graphite as a moderator, and water as a coolant.

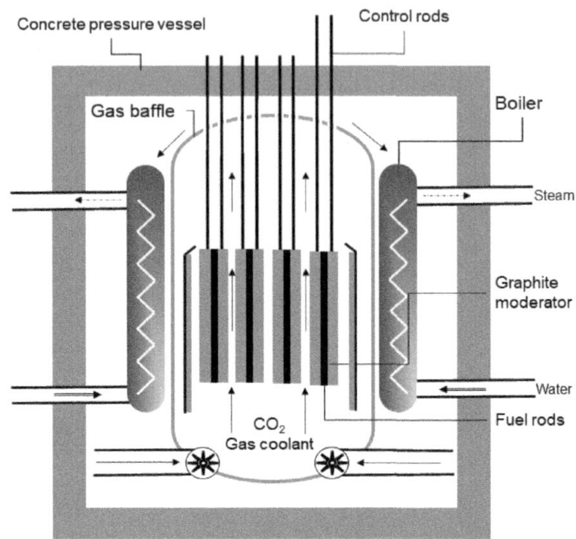

Figure 5.7. Schema of an AGR reactor. It uses enriched uranium as fuel, graphite as a moderator, and CO_2 gas as a coolant.

steam and water is directed to large separators installed above the core. The walls of the reactor cavity around the steel vessel are made of two meters thick reinforced concrete. Since 1990, RBMK fuel enrichment has risen from 1.8% to an average of 2.8%. It now includes about 0.6% erbium as a burnable absorber [7].

The advanced gas-cooled reactor (AGR) illustrated in Figure 5.7 was developed from the former Magnox reactor, which used natural uranium metal fuel, encased in magnesium alloy cans. Hence the name Magnox. Both Magnox and AGR reactors are basic gas-cooled reactors (GCR), exclusive to the UK.

AGR reactors use $^{235}UO_2$ pellets stacked in roughened stainless cans, which are clustered in three rings inside a graphite sleeve. These round fuel assemblies are loaded into the core vertical channels of graphite blocks [8]. Graphite blocks also provide channels for control rods and a structure to enhance the gas passing. The reactor is cooled by blowing carbon dioxide (CO_2) gas through the spaces between the fuel rod channels. This gas flows up and down through the core before going into the space above the gas baffle and into the gas-to-water boilers, where it transfers its heat to water to produce steam [9]. A concrete pressure vessel surrounds the reactor core and the boilers, maintaining the reactor coolant gas under pressure and providing radiation shielding [10].

Canadian deuterium uranium (CANDU) reactors (see Figure 5.8), or PHWR reactors, use natural UO_2 fuel bundles that are loaded into horizontal zirconium alloy pressure tubes. The pressurized heavy water

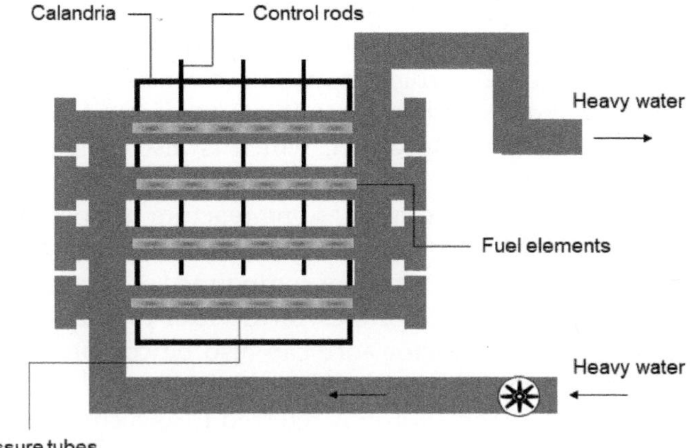

Figure 5.8. Schema of a CANDU reactor. It uses natural uranium as fuel and heavy water (D_2O) as a coolant and moderator.

(D_2O) coolant is pumped through the pressure tubes, over the fuel bundles. Hence, the reactor heart is the pressure tube, designed to contain the fuel bundle and the primary coolant [11]. The reactor vessel is the calandria,[6] a structure where the pressure tubes are immersed in the cold heavy water moderator. The pressure tubes pass through the calandria tubes separated from them by a gas gap to prevent heat loss. CANDU control rods are inserted or withdrawn in the calandria. After leaving the reactor core, the hot heavy water goes to the steam generator, where it is used to produce steam from normal water [12]. Apart from Canada, CANDU reactors are in operation in China, South Korea, Romania, India, Pakistan, and Argentina [13].

Below is a summary of some of the technical features of the thermal reactors depicted before [14]. In addition to the fuel, coolant, and moderator, the outlet pressure and temperature are indicated for each reactor type. Also, whether there is a secondary circuit for steam generation or not.

Type	Fuel	Coolant	MPa	°C	Moderator	Steam
PWR	3.5%–5% $^{235}UO_2$	Light water	16	300	Light water	Secondary circuit
VVER	2.4–4.4% $^{235}UO_2$	Light water	16	300	Light water	Secondary circuit
BWR	2.1–2.6% $^{235}UO_2$	Light water	7	280	Light water	Directly from boiling
RBMK	1.8–2.8% $^{235}UO_2$	Light water	6.9	280	Graphite	Directly from boiling
AGR*	2.5–3.5% $^{235}UO_2$	CO_2	2.7	400	Graphite	Secondary circuit
CANDU	Natural UO_2	D_2O	9.3	300	D_2O	Secondary circuit

* The modern type AGR may reach temperatures of up to 950°C.

---

[6] A horizontal cylindrical vessel of about 6 m long and 7 m across housing a matrix of horizontal tubes called calandria tubes. It contains the moderator at slightly above atmospheric pressure.

All mentioned reactors use control rods to keep up the state of the fission reaction by inserting or removing them from the reactor core. These are made of neutron-absorbing materials such as cadmium, hafnium, silver, indium, or boron. In the event of an unsafe condition, control rods are inserted for a rapid shutdown, which is also called an emergency shutdown or SCRAM.[7]

In addition to the control rods, some reactors use neutron-absorbing materials called poisons to reduce the excess reactivity, e.g., from fresh fuel. Burnable poisons, in the form of compounds of boron or gadolinium, are usually introduced as additives to the fuel to enable a longer life. Excess reactivity in the reactor core can also be compensated by boron dilution, i.e., by dissolving boric acid in the coolant or moderator. In the context of nuclear reactors, boron is referred to as a chemical shim and is used to produce spatially uniform neutron absorption [15].

The reactor containment structure, the last barrier of defense in depth, is a concrete and steel enclosure surrounding the reactor and primary circuit with walls at least one meter thick, able to withstand different accident scenarios including core melting. It provides confinement of radioactive materials, radiation shielding, and protection against external natural and human-induced events (earthquakes, plane crashes, pipe whipping, and so on) [16].

The containment structure is now designed with pressure-suppression systems to preserve its integrity and prevent the release of radioactive materials under growing pressure conditions, e.g., a LOCA accident. Typically, carwash-style sprayers attached to the upper walls are used to condense the steam and reduce the temperature and pressure inside the building [17].

PWR containment structure is the strongest because it must withstand higher internal pressures. Its dome design includes two pressure-suppression systems, that is, a spray system like the one mentioned above, and a fan cooler system The inside surfaces of the containment are covered with a steel liner to prevent any gas from escaping through cracks that may have developed in the concrete over time [18].

Containment designs for BWR reactors have evolved from Mark I to Mark II and Mark III, all relying on pressure suppression systems to limit the containment pressure. The water to condense the steam is stored in a suppression chamber (Mark I), a wetwell (Mark II), or a suppression pool

---

[7] According to Warren Nyer, an American physicist present at the time of the start-up of the first reactor Chicago Pile-1, Volney (Bill) Wilson is credited with coining the use of the word "scram" for the sudden shutdown of the reactor. He had recalled Wilson saying, "You scram out of here" when someone asked what to do if the red emergency button had to be pressed [NRC, Putting the Axe to the 'Scram' Myth].

(Mark III). The drywell is the primary containment structure. Its design allows it to contain and direct the steam that may result from a major pipe rupture. It also sends the steam and the drywell's air into the suppression chamber [19].

Mark I drywell has the shape of an inverted lightbulb with vent pipes directed to the toroidal[8] suppression chamber underneath. Mark II drywell is a truncated cone directly above the cylindrical wetwell. Mark III drywell is a cylindrical, reinforced concrete structure with a removable head, built over the suppression pool [20].

A schematic of the containment structure Mark I is shown in Figure 5.9. If a break occurs, steam will vent into the suppression chamber, through various descending pipes with open ends submerged in the water, filling about half the torus. The reactor building in BWRs is a secondary containment structure. It surrounds the drywell and is kept at a slight vacuum to control any radioactivity release.

CANDU reactors have two types of containment, single-pressure, and negative-pressure. Figure 5.10 shows a scheme of a single-pressure containment. It is used for large single units and consists of a domed prestressed concrete structure, an elevated dousing tank, dousing spray headers, airlocks, and a closure system. The concrete walls are over one

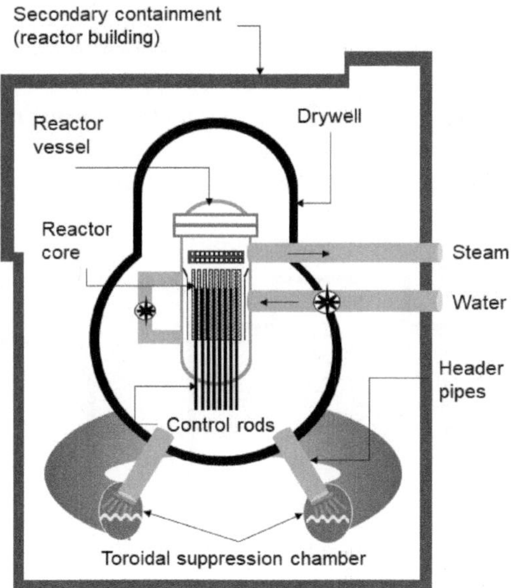

Figure 5.9. BWR Mark I containment diagram showing the reactor building—the secondary containment—and the drywell.

---

[8] A circular ring doughnut-shaped.

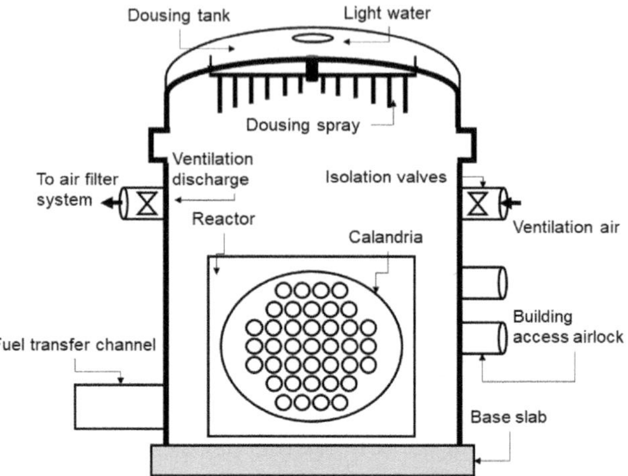

Figure 5.10. A simplified drawing of a CANDU single unit pressure suppression containment.

meter thick, and all internal surfaces, e.g., the upper dome, outer walls, and base slab are steel-lined for increased leak-tightness [21].

The negative pressure containment design is used for multi-unit CANDU stations, e.g., the Bruce Nuclear Generating Station which houses eight reactors.

A scheme of a vacuum building and one of the reactors connected to it is shown in Figure 5.11. In this containment system, a group of CANDU units, housed in separate reinforced concrete buildings, is connected to a single vacuum building by a large pressure relief channel labeled "vacuum duct" in Figure 5.11. If an increase in pressure or radioactivity

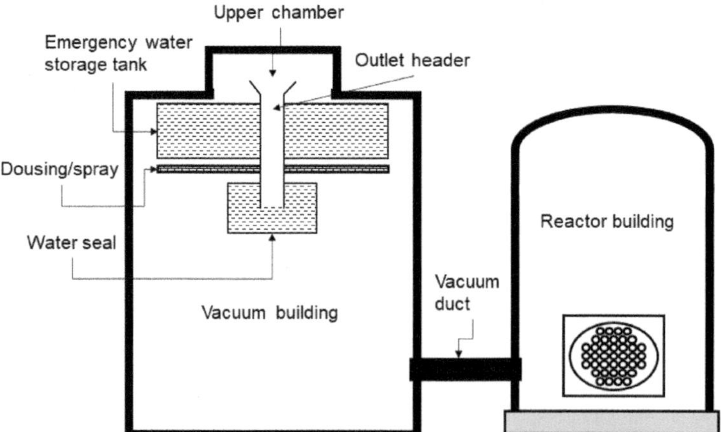

Figure 5.11. A negative pressure containment for CANDU reactors showing only one reactor and the vacuum building.

Figure 5.12. Bruce Nuclear Generating Station in Ontario, Canada (Photo courtesy of Chuck Szmurlo under Creative Common License CC BY 2.5).

level is detected in any of the units, the vacuum building rapidly draws the release via the relief channel and automatically closes all openings to isolate the reactor building. In case of an accident, all the units should be shut down.

Figure 5.12 is a photo of one of the plants of the Bruce Station with four reactors. It shows the four reactor buildings all interconnected during normal operation. The vacuum building can be noticed near the shore, on the left, in front of the four reactors connected to it.

British gas-cooled AGR reactors do not have a containment structure similar to light-water reactors. Instead, the reactor core—fuel and moderator—is in a domed steel vessel known as the gas baffle, and together with the boilers and pumps, is contained in a prestressed concrete pressure vessel with side walls usually 5–7 m thick that also provides shielding. See Figure 5.7 to have an idea. If needed, a secondary shutdown system injects nitrogen into the gas coolant to help stop the chain reaction [9].

In the event of a primary coolant pipe rupture, the gas (CO_2) will not undergo a sudden phase change like water does and the core will not become uncovered like in water-cooled reactors [22]. In addition, the hollow fuel pellets[9] in the fuel assemblies help keep the temperature at 650°C, while holding the released fission gases without increasing the internal gas pressure [23].

RBMK reactors neither have a containment structure similar to those in PWRs or BWRs. Instead, the reactor core is surrounded by a reinforced

---

[9] Temperatures in the center of solid fuel pellets can exceed 1000°C under the harsh conditions in which the fuel is operated.

concrete cavity that acts as a radiation shield, with heavy steel plates at the bottom and top, welded to the extensions of the fuel channels (see Figure 5.6). In addition, parts of the reactor and primary circuit are enclosed in individual containments designed to protect only that space [24].

Some milestones in advanced nuclear reactors

Because nuclear power pioneers believed in a possible scarcity of fissile material, back in the middle of the twentieth century, an experimental fast reactor was created in the US. The Experimental Breeder Reactor (EBR-I), demonstrated the prospect of breeding, i.e., the reactor capability of producing more fissile material than it consumes when hitting a fertile material with fast neutrons. As noted earlier, this reactor was also the first to generate usable electricity from nuclear energy in 1951, albeit just enough to feed four 200-watt light bulbs [25].

Fast reactors do not need a moderator to slow down neutrons, but they do require more enriched fuel than thermal reactors. However, the EBR-I used fully enriched uranium metal rods as fuel and blanket rods of natural uranium surrounding the core to breed fissile ^{239}Pu. This experiment proved the possibility of transmuting ^{238}U into ^{239}Pu. It has also demonstrated the viability of producing controlled heat in a reactor without a moderator and using liquid metal as a coolant [26, 27]. The EBR-I was in operation for 12 years.

In 1964, a second breeder reactor, the EBR-II, was operating at 30 MWt (thermal). It proved to be a success in demonstrating the feasibility of a complete breeder-reactor power facility with an *on-site* fuel cycle. The fuel cycle was meant to reprocess the burned metallic fuel, produce subassemblies, and assemble control and safety rods [25]. Instead of sodium-potassium alloy, the EBR-II employed molten sodium as a coolant, a pool-type design like the one shown in Figure 5.13, and different types of fuel.

As can be seen in Figure 5.13, a pool-type design is where the reactor core, its primary heat exchanger, and the fuel-handling equipment are immersed in molten sodium. This made the EBR-II passively safe, implying that it could safely shut down, without operator assistance, even if all safety systems had failed. This point was confirmed in 1986, in experiments carried out at full reactor power, and with the automatic shutdown features intentionally disabled [28].

In the mid-1980s, the EBR-II played an important role in the Argonne National Laboratory integral fast reactor (IFR) project to develop the technology for a system using a sodium reactor, an entire fuel cycle, and waste management technologies. Two tests simulating accidents involving loss of coolant flow demonstrated the inherent safety of this reactor. Even

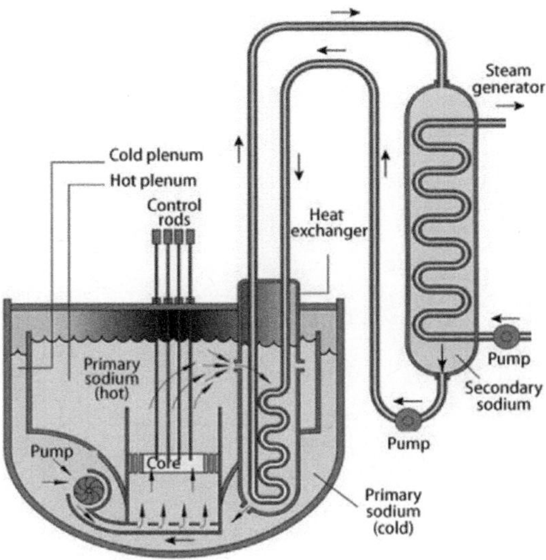

Figure 5.13. Pool-type sodium-cooled fast reactor (Image courtesy of the US DOE).

when the normal shutdown devices were disabled, the reactor shut down safely on its own. The project ended in 1994 for reasons unrelated to science and technology [29].

Russia commissioned the first commercial prototype of a fast neutron reactor in 1973, in Aktau, Kazakhstan. The BN-350 was a sodium-cooled reactor with a loop-type[10] design and a capacity of 350 MWe. It was fueled with uranium oxide enriched to 17–26% and produced electricity and desalinated seawater at the same time. The reactor operated for 27 years until Kazakhstan decommissioned it in 1999 [30]. The BN-350 was constructed based on the methods and materials tested in the BOR-60 program, a sodium-cooled fast reactor test facility installed at the Research Institute of Atomic Reactors, in Dimitrovgrad, Russia, in 1964 [31].

Russia's next step was a commercial-size sodium-cooled fast reactor, the BN-600 (600 MWe), followed by the BN-800 (880 MWe). The BN-600 is a pool-type design breeder reactor that uses uranium oxide enriched to 17–26%. It was connected to the grid in 1982 and has been operating since then. The BN-800, shown in Figure 5.14, is a three-loop pool-type reactor of similar overall size and configuration to the BN-600. It is used as a burner reactor and has been in commercial operation at the Beloyarsk nuclear power plant since it was connected to the grid in 2015 [32].

---

[10] In opposition to the pool type reactor, in the loop type reactors, the heat exchangers are outside the reactor vessel.

Figure 5.14. BN-800 sodium-cooled fast reactor at Beloyarsk nuclear power plant in Russia (Creative Commons photo courtesy of Rosatom).

The BN-800 shows some significant enhancements compared to the BN-600. Since one of its objectives was to demonstrate the performance of a closed fuel cycle, it includes fuel flexibility, varying from U+Pu nitride, MOX (UO_2/PuO_2), or metal fuel. Also, certain safety features have been added. For example, a new passive emergency shutdown system, a sodium cavity over the reactor core to reduce sodium void reactivity in case sodium boils in the core; and a core catcher at the bottom of the reactor vessel to prevent debris scattering in the event of a severe accident [33].

The BN-800 was first loaded with enriched uranium fuel and partially with MOX fuel bundles, but that fuel was fully replaced with fresh MOX in 2022 [34]. The BN-800 closed fuel cycle means that the produced plutonium and minor actinides are consumed when this reactor is operating.

The French experience with fast reactors focused on three reactors, Rapsodie from 1967 to 1983, Phénix from 1973 to 2003, and Superphénix from 1985 to 1997. Rapsodie was an experimental loop-type sodium-cooled reactor with a nominal capacity of 20 MWt, fueled with a mix of PuO_2 and UO_2. It was used for 15 years in irradiation tests on fuels and materials.

Phénix was a pool-type sodium-cooled reactor of around 250 MWe. It was connected to the grid in 1973. In 1989, it began experiencing several unexplained reactivity transients, along with some sodium leaks and fires that affected its load factor and safety, forcing its shutdown and major renovations. It was restarted in 2003 with decreased nominal power (133 MWe) and was used for the research and development of sodium-cooled fast reactors in areas such as core physics, safety, and technology innovations. All the plutonium produced was recycled after reprocessing. Phénix was permanently shut down in 2009, after 35 years of operation [35].

Superphénix was a 1200 MWe fast breeder prototype reactor, sodium-cooled, with a pool-type design. It used MOX as fuel and depleted uranium oxide blankets for breeding. Superphénix was connected to the grid for 12 years, from 1986 to 1998, but it only operated for 53 months. The rest was spent on reasonable technical modifications and unnecessary administrative delays. A final decision was made in 1998, for political reasons [36]. Several events arose in those twelve years, but only a few can be classified as incidents or anomalies consistent with the International Nuclear and Radiological Event Scale (INES) [37].

The Prototype Fast Reactor (PFR) at Dounreay, UK, operated from 1975 to 1994. It was a commercial-size reactor using sodium as a coolant and $(Pu+U)O_2$ as fuel, designed to produce 250 MWe. During its operation, it provided valuable experience as a pool-type fast reactor. It also served as a test bed for fuel assemblies, components, and materials [38]. Numerous cladding materials were studied, including austenitic steels, ferritic steels, and Nimonic alloys.[11] An important achievement at the time was the detection of gas-space leaks in the steam generators and the replacement of the welds with sleeves to end them [39]. Considerable experience in decommissioning was also obtained after the shutdown.

Monju was the second Japanese prototype of a sodium-cooled fast-breeder reactor. It was a loop-type reactor of 280 MWe, fueled with MOX, which achieved criticality in 1994. Before Monju, the experimental fast reactor Joyo ran for more than 26 years. Joyo went critical in 1977 and needed several core changes to increase its power [40]. Unlike Joyo, Monjo was shut down after a leak and a fire in 1995 and needed to be subjected to a series of improvements in various systems. Its restart was finally authorized in 2010, but it was suspended again after a fuel exchange device accidentally fell into the reactor. The permanent shutdown of Monju occurred in 2016.

Despite being a developing country, India has pursued an indigenous nuclear power program aiming at exploiting its own uranium and thorium resources to produce electricity, thanks to the legacy of Dr. Homi Bhabha (1909–1966), a great physicist who was the founder of the Indian nuclear energy program. This pursuit has led to significant progress in areas such as uranium exploration and mining, fuel fabrication, heavy water production, reactor design and construction, and waste management. In

---

[11] Nimonic is a registered mark for a family of nickel-based high-temperature low-creep superalloys.

1974, India conducted a nuclear test which led to international sanctions isolating it from the rest of the world's nuclear technology. India is not a party to the NPT but, since an agreement with the US in 2005, has chosen to sign safeguards with the IAEA[12] for its civil nuclear facilities [41, 42].

The program devised by Dr. Bhabha in the 1950s covers a three-stage thorium plan aimed at the utilization of India's vast thorium reserves. The first stage included the deployment of heavy water reactors (PHWR), using the native natural uranium to produce electricity and, ^{239}Pu from the spent fuel reprocessing. The second stage consisted of fueling a fast breeder reactor (FBR) with ^{239}Pu to produce electricity and convert ^{232}Th blankets into ^{233}U. The third stage was based on a 300 MWe Advanced Heavy Water Reactor (AHWR), and a ^{232}Th-^{233}U close cycle. India has already been experimenting with burning ^{232}Th in its indigenous heavy water reactors [43].

Currently, India has 19 PHWRs in operation [44] and three reprocessing plants [45]. Also, a sodium-cooled Fast Breeder Test Reactor (FBTR) in operation since 1985. It uses a mixed core of Pu+U carbides to gain experience in the design, construction, and operation of fast breeder reactors. Furthermore, since 1997, India has been carrying out studies in a unique research reactor called Kamini (Kalpakkam mini reactor), in which the fuel is ^{233}U, in the form of a uranium aluminum metal alloy, obtained from irradiated ^{232}Th [46].

India's first 500 MWe Prototype Fast Breeder Reactor (PFBR), the second stage of its nuclear program, started construction in 2004 at Kalpakkam. Although its schedule has been delayed repeatedly, it is almost ready to be commissioned [47]. India's PFBR has a pool-type design and uses MOX fuel. Blanket assemblies with fertile ^{232}Th and ^{238}U could be placed around the fuel core to obtain fissile ^{233}U and ^{239}Pu as well. A Fast Reactor Fuel Cycle Facility (FRFCF) is also planned to reprocess the spent fuel and close the fuel cycle for both MOX and mixed carbide fuels [45].

Advanced nuclear fuel development

New concepts are emerging to improve the performance, safety, and proliferation resistance of existing nuclear reactors alongside new advanced reactors. These include novel advanced accident-tolerant fuels, fuel that aim at higher burnup, and the high-assay low-enriched uranium (HALEU) fuel.

---

[12] India signed an INFCIRC/66-type safeguards agreement (INFCIRC/754) that entered into force in 2009, and an Additional protocol that covers 31 nuclear facilities, the INFCIRC/754 Agreement between the Government of India and the International Atomic Energy Agency for the Application of Safeguards to Civilian Nuclear Facilities, 2009 and Addendum 12.

An accident-tolerant fuel (ATF) is described as a nuclear fuel more tolerant to severe[13] accident conditions. The term is derived from the experience of the Fukushima nuclear accident, which signaled certain vulnerabilities of current fuels. In line with the NRC definition, the term identifies a set of new technologies that have the potential to enhance nuclear power plant safety by offering better performance during normal operation, transient conditions, and accident scenarios [48]. According to the Department of Energy, these new technologies can be in the form of advanced cladding and fuel pellet designs [49].

To reduce hydrogen buildup, improve fission product retention, and achieve a fuel that is structurally more resistant to radiation, corrosion, and high temperatures in comparison with the current standard UO_2–Zircaloy system, new technologies focus, e.g., on metal fuel forms rather than ceramic oxides, and on improving pellet-cladding interaction, fission gas retention, and oxidation resistance.

Some examples of innovative fuel cladding and new fuel pellet designs are the chromium coating fuel rods combined with chromia-doped[14] fuel pellets from Framatome, in operation at the Calvert Cliffs NPP [50] since 2021; the Fe-Cr-Al steel alloy cladding for conventional UO_2 fuel from GNF (Global Nuclear Fuel) being tested in the Clinton NPP [51]; as well as the chromium-coated cladding and the advanced doped UO_2 fuel, from Westinghouse, that are being tested in Station 2 of Byron NPP [52].

Another example is the Lightbridge Fuel™ illustrated in Figure 5.15. It is a novel extruded metal fuel developed by Lightbridge Corporation; a company based in Virginia. This proprietary fuel, under design basis accidents,[15] does not generate hydrogen or release fission products, thus enhancing safety. In the helical multi-lobed design seen on the right corner of the illustration, the zirconium-niobium cladding is metallurgically bonded to the fuel, hence removing the source of fission product release in case of a barrier breach and ensuring expansion during burnup to accommodate swelling [53].

The uranium-zirconium alloy core of the Lightbridge fuel not only improves thermal conductivity compared to UO_2 ceramic fuel, but its cruciform geometry also contributes to a cooler average fuel operating temperature of up to 360°C.

---

[13] A severe accident is a very unlikely event brought about by multiple failures which is beyond the design basis accident and can cause significant core degradation.

[14] Chromium (III) oxide (Cr_2O_3).

[15] A design-basis accident is a postulated accident that a nuclear facility must be designed and built to withstand without loss to the systems, structures, and components necessary to ensure public health and safety (NRC definition).

Figure 5.15. Helically-twisted Lightbridge Fuel™ test assembly (Image courtesy of Lightbridge Corporation).

Considering the readiness of the concepts proposed so far, these have been categorized into near-term and long-term. The concepts with the highest technological readiness at this time are in the near term. Amid them are [48]:

- chromium coating of zirconium alloy claddings, and FeCrAl cladding,
- doped UO_2 pellets,
- increased fuel enrichment by up to 10%, and
- increased burnup limit to 75 or 80 GWd/t.

Because zirconium alloy cladding can react with the water vapor formed in the event of severe accidents and release extremely flammable and explosive hydrogen gas, nuclear fuel fabricators like Westinghouse, Framatome, and Global Nuclear Fuel (GNF) are currently testing various zirconium alloy claddings coated with a thin layer of metallic and ceramic materials. Experiments conducted in research reactors at the Massachusetts Institute of Technology, Halden Institute for Energy Technology, and ORNL High Flux Isotope Reactor in the presence of steam and temperatures from 1000 to 1200°C have demonstrated that such coating reduces zirconium cladding oxidation following a LOCA accident, and to some extent, beyond it [54].

FeCrAl alloys also seem to be compatible with current standard UO_2 fuel, possess better corrosion resistance, and produce less hydrogen in a LOCA scenario than zirconium alloys [55]. The FeCrAl-based alloy named IronClad shown in Figure 5.16, was developed by Oak Ridge National Labs and Global Nuclear Fuel.

Adding a dopant, e.g., chromium (Cr_2O_3) and aluminum (Al_2O_3), into UO_2 pellets during manufacturing can modify some physical properties and reduce fuel degradation and cladding damage. The NRC has already approved doped pellets for accident-tolerant fuels for BWR from Framatome and Global Nuclear Fuel [48], and for PWR from Westinghouse [56].

Figure 5.16. IronClad, an iron-chromium-aluminum alloy developed as an alternative to zirconium alloy fuel cladding for accident-tolerant fuels (Image courtesy of U.S. Department of Energy).

Figure 5.17. Transient Reactor Test Facility (TREAT) at the Idaho National Laboratory (INL) (Photo courtesy of the US Department of Energy).

Accident-tolerant fuel technology testing in commercial reactors started in 2018. Different lead test rods and assemblies[16] have been inserted for irradiation tests at Byron (PWR), Calvert Cliffs (PWR), Vogtle (PWR), Brunswick (BWR), Clinton (BWR), and Hatch (BWR) nuclear power plants [57].

An important facility for the testing of advanced fuels, unique in the US, is the Transient Reactor Test Facility (TREAT) at the DOE's Idaho National Laboratory (INL), shown in Figure 5.17. TREAT is a thermal

---

[16] These are fuel rods and assemblies with design features or materials that have not been approved by the NRC for unrestricted use in the reactor core.

air-cooled, graphite-moderated reactor, designed to test fuels and structural materials. It is capable of producing short-term bursts of intense neutron flux to simulate accident conditions. In TREAT, the sample behavior can be monitored in real-time and the experiment can be filmed [58].

An increase in the burnup limit of up to 75 or 80 GWd/t, together with a consistent increase in ^{235}U enrichment to 10% is also a prospect to explore in the near term. In PWRs, these changes are likely to increase the refueling cycles while reducing refueling outages. It will also be possible to reduce the amount of high-level radioactive waste intended for long-term storage and disposal. Modifications are to be expected in how fuel storage and transportation are done and regulated today [59].

Long-term accident-tolerant fuel concepts are those with more uncertainties that require even more data and the development of new models and methods to be commercialized. These include fuels like uranium nitride (UN) pellets and the above-mentioned extruded metal, as well as various silicon carbide (SiC) composite materials for cladding [48]. UN pellets were selected over silicide (U_3Si_2) and carbide (UC) pellets; and SiC composite materials over other cladding materials, after various experiments, tests, and simulations proved their suitability [60]. SiC composite cladding is made up of SiC woven fibers embedded in a matrix of the same material to form a rigid tube, frequently referred to as composite SiC_f/SiC_m [61].

Advanced fuels include TRISO fuel, developed for the first time for gas reactors in the UK, Germany, USA, Japan, South Africa, and China for an exceptional experience. The tri-structural isotropic (TRISO) particle fuel, consisting of tens of thousands of dispersed particles of nuclear fuel embedded in a dense matrix, is considered the most robust nuclear fuel known today. It is more resistant to irradiation, oxidation, and corrosion, also providing higher burnup and high stability up to 1600°C [62]. Fissile material in TRISO can be uranium oxide (UO_2), uranium nitride (UN), uranium oxycarbide (UCO)—a solid-state solution of uranium dioxide and uranium carbide, uranium dicarbide (UC_2), and even it may be thorium-based [63].

As illustrated in Figure 5.18, TRISO fuel is fabricated in various geometric forms, including cylindrical compacts, also called pellets, or spheres the size of a billiard ball, called pebbles.

Figure 5.19 shows the internal layers surrounding the small fuel kernel in a TRISO particle, which allow each particle to act as its containment. The first layer from inside to outside is a porous carbon buffer that provides space for gas accumulation. Then comes an inner layer of pyrolytic carbon[17]

---

[17] Pyrolytic graphite is a synthetic form of graphite that is deposited on a suitable underlying substrate from the vapor phase by thermal decomposition of a simple hydrocarbon such as methane.

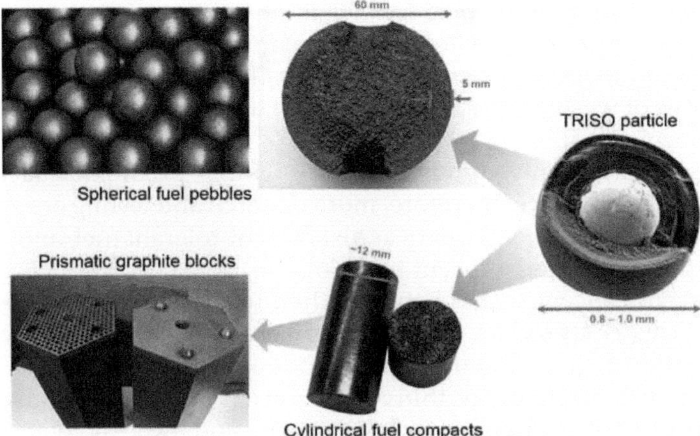

Figure 5.18. Tri-structural isotropic (TRISO) particle fuel forms. Spherical pebbles are shown above, and cylindrical compacts used in prismatic blocks are below. The coated TRISO particle is shown on the right (Image courtesy of the US NRC).

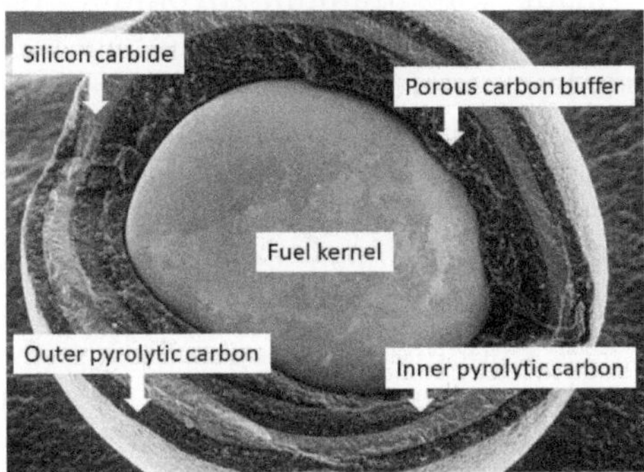

Figure 5.19. Enlarged tri-structural isotropic (TRISO) kernel to show its internal layers of carbon and carbide that serve as the primary containment for radioactive material. (Image courtesy of the US Department of Energy).

which provides a surface for the deposition of the silicon carbide and adds to the retention of nongaseous fission products. The silicon carbide (SiC) layer that follows—the third—is the one for structural strength and is the most important in the retention of nongaseous fission products. The outer pyrolytic carbon layer protects the SiC and bonds the particle to the cylindrical pellet or pebble. If the particle is for utilization as deep-burn TRISO, SiC is replaced by zirconium carbide (ZrC) [64].

TRISO fuel is structurally very resistant to high temperatures, corrosion, oxidation, and neutron irradiation. It cannot melt in a reactor and offers a sturdy barrier that prevents the release of radioactive fission products. In addition, TRISO can achieve higher burnup—three times the burnup that current light-water fuels can accomplish—and reduces enrichment and proliferation risk. It is also supposed to be good for long-term storage in repositories. Some corrosion measures have shown that it should maintain its integrity for a million years or more. However, the volume of waste from spent TRISO fuel could be very high despite the higher burnup [65].

Presently, the only commercial-scale TRISO production facilities in the world are in the US. Therefore, more research and testing are advisable to assess its feasibility and safety as an accident-tolerant fuel, and, also, for its future management as spent fuel [66].

A recent concept that is being studied for LWRs is the Fuel-in-Fiber (F-in-F). This is a product of Free Form Fibers [67], a commercial entity using chemical vapor deposition to produce high-performance ceramic fibers. Using this technology, thin cylindrical UN fuel fibers are surrounded by layers of porous carbon, pyrolytic carbon, and SiC. All layers are then wrapped in a SiC external coating, similar to TRISO but maximizing the uranium load. The fibers are finally embedded in a SiC matrix to form a pellet the size of a regular LWR fuel pellet [68].

Metal fuels made of uranium-zirconium (U-Zr) alloy and uranium-plutonium-zirconium (U-Pu-Zr) alloy are best for fast reactors, among other reasons, because of their superior thermal conductivity, higher densities of fissile and fertile materials, enhanced breeding potential, and lower fissile inventory compared to oxide fuels. They have also recently gained attention in the search for higher burnups and accident-tolerant fuels. Major concerns have been their accelerated swelling and pellet-cladding interactions from irradiation in normal operation and their risk of generating large amounts of hydrogen during severe accidents [52].

A program for the accelerated promotion of advanced metal fuels from the early 2020s has already produced valuable results including the use of an annular geometry for the fuel; additives to chemically bind the lanthanides; and advanced alloys of molybdenum, titanium, and zirconium to improve fuel properties [69]. More recently, great results have been obtained with a fuel that is extruded and metallurgically bounded to the cladding, like the Lightbridge Fuel™ shown in Figure 5.15. Metal fuels are primarily designed for advanced reactors, hence higher uranium enrichments, above the anticipated 10%, appear to be needed.

Liquid fuels are also an option for advanced reactors, typically for molten-salt fast reactors (MSR). These reactors use liquid salts as a coõlant, usually fluorides or chlorides, in which the fuel can be dissolved. For example, in the 1960s molten salt reactor experiment at Oak Ridge National Laboratory, the low-enriched $^{235}UF_4$ was dissolved in the molten mixture of lithium and beryllium fluoride used as a coolant [70].

With liquid fuels, there is no fuel to fabricate, replace, store, or dispose of. Neither online refueling is necessary. Natural and depleted uranium, as well as spent fuel, can be used for operation. Presently, leading energy companies in the US are partnering to develop a zero-power molten chloride fast reactor (MCFR) that will use liquid salts as both a coolant and fuel [71].

Another fuel in development, aimed to improve the performance, economics, and safety of CANDU and other pressurized heavy water reactors, is ANEEL. It is a proprietary fuel technology that uses thorium and high-assay low-enriched uranium (HALEU). ANEEL is expected to reduce the amount of waste produced by 80% and also to provide non-proliferation benefits [72].

Generation IV nuclear reactors

The average life of a nuclear power plant worldwide is currently approaching 50 years. In the US, the country with the largest reactor fleet in the world, 83 out of 92 operating reactors are between 30 and 50 years old, and 6 already surpassed 50 years, leaving only a few reactors with less than 30 years [73]. These light and heavy water reactors mostly have served and still are serving the world well.

Global efforts to achieve carbon neutrality and climate change mitigation have generated renewed interest in nuclear energy, along with wind, solar, and energy storage energy.[18] Many new and abandoned technologies have been explored over the past decade to make nuclear reactors even safer, more flexible, less costly, and versatile, as well as more proliferation-resistant. The aim is also to generate less nuclear waste.

Generation IV reactors include nuclear systems that are expected to increase efficiency, safety, and reduce radioactive waste. They could also support desalination, and hydrogen production, and provide heat for other industries [74].

The Gen IV International Forum (GIF)[19] has selected six reactor technologies for further development. They are [75]:

- gas-cooled fast reactor (GFR),
- lead-cooled fast reactor (LFR),
- molten salt reactor (MSR),
- sodium-cooled fast reactor (SFR),
- supercritical-water-cooled reactor (SCWR), and
- very high-temperature reactor (VHTR).

---

[18] Energy storage is a technology that holds energy produced at one time for use at a later time.

[19] The Gen IV International Forum is a cooperative international endeavor seeking to develop the research necessary to test the feasibility and performance of advanced nuclear systems, and to make them available for industrial deployment by 2030.

To be more efficient than conventional reactors, advanced reactors would have to run at higher temperatures and much longer without refueling, thereby reducing downtime. For example, the output temperatures of the gas-cooled reactor are between 750°C and 950°C, allowing power conversion efficiencies of up to 48% [76].

Likewise, the International Atomic Energy Agency (IAEA) Advanced Reactors Information System (ARIS) database defines the following advanced nuclear systems [77]:

- water-cooled technology
- gas-cooled technology
- molten-metal-cooled technology
- molten-salt-cooled technology
- small and medium reactors

Advanced nuclear systems, i.e., the new generation of reactors, include both evolutionary (Generation III+ water-cooled technology) and innovative technologies (gas-cooled, liquid metal, and molten salt), as well as small modular reactors (SMR). Evolutionary concepts are designs that enhance existing technologies through specific modifications while maintaining proven features to minimize technological risk. Innovative designs, by contrast, incorporate radical changes. SMRs are based on either existing technologies or advanced reactor designs.

Some data on the reactors under development is shown in Table 5.1 [77]. Each reactor type is broken down according to its objective, namely commercial, demonstration or prototype, and experimental. The table

Table 5.1. Reactors in development by type, purpose, and stage.

Reactor type	Purpose			Stage					TOTAL
	Comm.	Dem.	Exp.	Design*	Const	Operat	Rev/Lic**	On hold	
BWR	5	1		4		1	1		6
GCR	3	2		3	1			1	5
GFR	1	2		3					3
HWR	5			3		2			5
iPWR	3			1			2		3
LFR	4	5	3	12					12
MSR	6	2		8					8
PWR	22			13	5	2	1	1	22
SCWR	3			3					3
SFR	6	3	1	9	1				10
SMR	33	11	4	42	2	1	2	1	48

*   Conceptual design, design, or detailed design
** In review or licensed

also indicates how many of each type are under design, construction, operation, and review by the regulatory authority, licensed, or on hold.

Reactors currently being designed or built for short-term commercial use are mainly PWRs or Small Modular Reactors (SMR) type. Water-cooled small modular reactors only differ from traditional reactors in size, simplicity, and efficiency. Innovations in water-cooled reactor designs involve changes in fuel and cladding materials, as mentioned earlier, and particularly in safety.

Table 5.2 lists several advanced evolutionary reactors that were formerly classified in Generation III+. It includes its electrical capacity, the country where it is operated or built, its status by country, the type of fuel it uses, and the current or expected efficiency[20] and burnup [78].

Table 5.2 also summarizes the most relevant enhanced safety features introduced into the reactor design. These include the use of passive safety systems, i.e., actioned by natural forces like pressurized gas, gravity flow, natural circulation flow, and convection. Most of these reactors have included a core catcher system to retain the molten core in the vessel in the event of a severe accident with core melting. Passive autocatalytic containment recombiners have also been installed to remove hydrogen if formed, and prevent hydrogen explosions.

Evolutionary reactors have already been installed in the United States, China, South Korea, the United Arab Emirates (UAE), Finland, Russia, Belarus, India, and Pakistan. The construction of the APWR designed by Mitsubishi was deferred in Japan, and its safety review by the NRC was suspended due to the policy adopted by the government after the Fukushima accident. Nonetheless, the EU APWR received a certificate of compliance with the European Utility Requirements (EUR) in 2014.

Regarding heavy water reactors, India has two IPHWR-700 under construction, completing a fleet of domestically-designed reactors. Canada, however, deferred two of its enhanced CANDU 6 (EC6) on account of demand uncertainties.

Traditional reactor safety systems strongly rely on redundant electrical and mechanical devices to act in the event of an accident. As a lesson from the Fukushima accident, where the lack of backup electrical power severely affected the cooling function of the reactors, special attention is now being given to passive systems in combination with active systems. Passive systems act on natural forces such as gravity, pressure differences, or natural heat convection [79]. Other safety improvements aimed at severe

---

[20] The efficiency of a nuclear reactor is the amount of electric power produced for each unit of thermal power. Typical efficiencies are around 33–37%, comparable to fossil fueled power plants.

Table 5.2. Advanced evolutionary, large, water-cooled reactor designs (as of April 2023).

Reactor		Capacity MWe	Country	Status	Fuel	Efficiency %	Burnup GWd/t	Safety
Type	Model							
PWR	AP1000	1100	4 China 2 USA	Operating Operating and final testing	$^{235}UO_2$	32	60	Active + passive Core catcher Hydrogen recombiner
PWR	APR1400	1400	3 S. Korea 2 S. Korea 2 UAE 2 UAE	Operating Construction Operating Construction	$^{235}UO_2$	35	> 55	Active + passive
PWR	APWR	1500-1600	3 Japan	**Deferred**	$^{235}UO_2$ Pl MOX		> 55	Active + passive + inherent
PWR	US-APWR	1700		**Suspended**	Pl MOX	37	62	
PWR	EU-APWR	1700		EUR certification	$^{235}UO_2$ Pl MOX	37	62	
PWR	ATMEA1	1200	4 Turkey	**Abandoned**	$^{235}UO_2$ MOX	37		Active + passive as backup Core catcher
PWR	EPR	1650	2 China 1 Finland 1 France 2 UK	Operation Operation Pre-operation Construction	$^{235}UO_2$ MOX	37	65	Active + passive Core catcher
PWR	HPR1000 (Hualong One)	1170	2 China 2 Pakistan	Operation Operation	$^{235}UO_2$	36		Active + passive as backup able to operate x 72 h
PWR	VVER 1200	1200	2 Russia 2 Belarus 3 Egypt 4 Turkey	Operation Operation Construction Construction	$^{235}UO_2$	34	60-70	Active + passive

BWR	ABWR	1350	4 Japan 2 Taiwan 1 Japan	Operation **Halted** **Suspended**	$^{235}UO_2$ Full MOX		50	Modularized design Automation
BWR	ESBWR	1560	2 USA	NRC approved. **Construction deferred**	$^{235}UO_2$	34	55	Passive + natural circulation No AC needed x 7 days Core catcher
BWR	KERENA	1250	1 Finland	Abandoned EUR approved	$^{235}UO_2$ MOX	37	65	Active + passive
HWR	EC6	740	2 Canada	**Deferred**	Flexible*			Active + passive
HWR	IPHWR-700	700	4 India	1 Operation 1 Testing 2 Construction	Nat. UO2 Flexible*		7 15	Active + passive

* Includes slightly enriched uranium, high burnup MOX, and thorium, in addition to natural uranium oxide

accidents in advanced water reactors include a core catcher for corium[21] retention, and hydrogen catalytic recombiners to prevent hydrogen explosions [80].

Innovative water-cooled reactors include more advanced technologies such as the supercritical water reactor (SCWR), as well as the integral power water reactor (iPWR).

Supercritical water reactors are considered to have a higher thermal efficiency than most LWRs because they operate at the thermodynamic critical point of water—pressure 22.1 MPa and temperature 374°C. In SCWRs, there is no change in the phase of water in the core, and consequently, high thermodynamic efficiency and plant simplification could potentially be expected. The concept is based on the successful operation of supercritical thermal power plants. Various conceptual designs have been proposed in Canada, Japan, and Russia, along with the Chinese CSR 1000 design and the HPLWR European design. These designs are technically feasible, but thus far they lack performance and operability confirmation [81].

The integral power reactor (iPWR) is a small modular reactor whose primary circuit components are located inside the reactor pressure vessel, thereby removing all the primary circuit pipework. iPWR design is intended to minimize the likelihood of loss of coolant accidents (LOCA), thus improving the safety and reliability of the reactor. There are several iPWR models from China, Argentina, Japan, South Korea, Russia, and the US.

The US NuScale power module, based on proven light-water reactor technology, with a power output of 50 MWe, is an example of an iPWR. The NRC issued the standard design certification for the NuScale, the first ever for a small modular reactor, in 2022 [82]. By design, the reactor building can hold up to 12 NuScale modules fully-factory built. The NuScale containment vessel is almost entirely submerged in a below-grade water pool, which provides passive cooling in an emergency and is also its heat sink.

As seen in Figure 5.20, the NuScale containment vessel is an upright cylinder closed at the top and bottom, and the compact steel reactor vessel inside it houses the core, pressurizer, and two steam generators. Natural convection and gravity are used to passively cool the reactor without additional water, power, or operator action. The fuel is less than 4.95% $^{235}UO_2$ in a core configuration of 37 fuel assemblies.

---

[21] A lava-like mixture of nuclear fuel, fission products, and reactor parts that can be created during a nuclear meltdown accident.

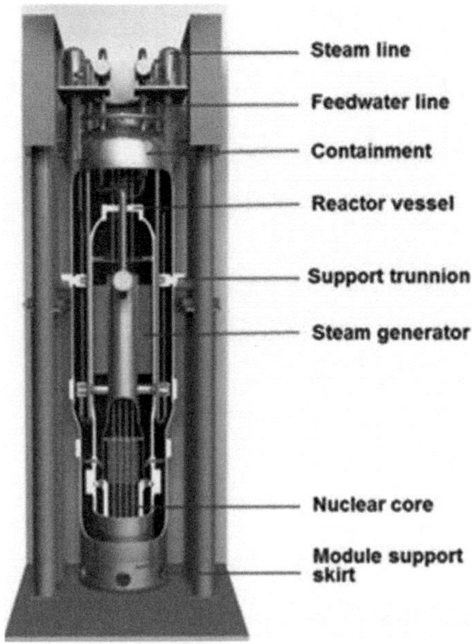

Figure 5.20. NuScale SMR module (Courtesy of the NRC).

Other iPWR small reactor models in advanced development are:

SMART	A 100 MWe iPWR small modular reactor developed by the Korea Atomic Energy Research Institute (KAERI), which received standard design approval from the South Korean regulator in 2012.
RITM-200M	A Russian 50 MWe iPWR small modular reactor that is a floating nuclear power unit based on experience with icebreakers. Currently, two floating nuclear power plants are under construction. They are expected to power Cape Nagloynyn, Chukotka, in the Russian Arctic.
RITM-200N	A 55 MWe iPWR small modular rector, similar to the RITM-200M but for onshore installation. In 2023, it was licensed for construction in the Republic of Sakha (Yakutia), in the Russian Arctic.
CAREM25	The prototype of 27 MWe iPWR from Argentina. It has been under construction since 2014.
ACP100	A 125 MWe iPWR (also called Linglong One) demonstration plant that is part of China's efforts in the area of SMRs and can be used for remote areas, industrial applications, or district heating. Its construction started in 2021.

Two more small modular reactors are now in the NRC pre-application review. They are the SMR-160, a 160 MWe iPWR that incorporates passive safety systems, and the BWRX-300™, a 300 MWe boiling water SMR with natural circulation and passive safety systems, based on the ESBWR large reactor (see Table 5.2) whose design is already certified by the NRC.

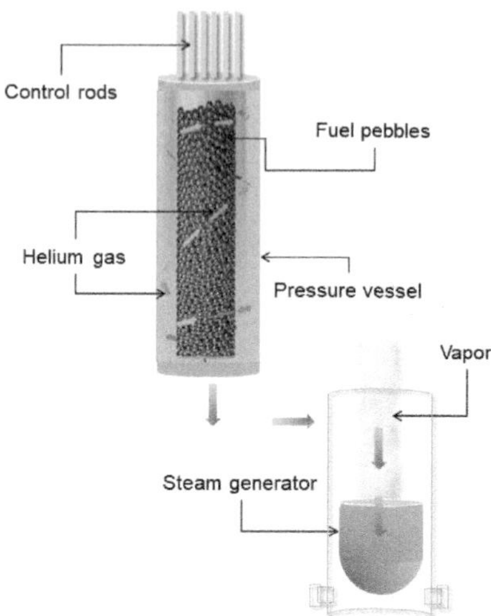

Figure 5.21. Simplified schema of the high-temperature gas-cooled SMR Xe-100.

The BWRX-300 is the first to complete two phases of the Vendor Design Review process in Canada and was selected by Ontario Power Generation to be deployed at its Darlington nuclear site. There is also interest in the BWRX-300 in Estonia, Poland, and in the US [83].

Only about 30% of the advanced reactors currently under development use non-water coolants. Non-water coolants include gas, molten metal, or molten salt. Except for the few large reactors already depicted, most of these are SMRs.

The HTGR (high-temperature gas-cooled reactor) is one of the advanced reactor technologies that combines high operating temperatures with varying pressures. It uses pure graphite as a moderator for a thermal reactor or no moderator for a fast neutron reactor. The preferred coolant is helium-inert gas, although CO_2 has also been considered. An example of an HTGR is the Xe-100 whose schematic is shown in Figure 5.21.

The Xe-100 is an 80 MWe pebble-bed SMR that uses ≈ 220,000 TRISO pebbles of 15.5% enriched uranium oxycarbide (UOC) as fuel and helium gas as a coolant. The pebbles are located in the cylindrical cavity formed by the graphite block reflectors inside the reactor vessel. The Xe-100 reactor operates at high temperatures, features a highly integrated protection system, and provides a high burnup of 160 GWd/t by multi-pass recycling of pebbles. It is claimed that the pebble bed cannot melt down because the heat generation is controlled as a result of the system's negative temperature coefficient [84].

The helium-cooled SC HTGR, and two fast modular reactors (FMRs), are also among the high-temperature reactors that are in development. The SC HTGR is a 272 MWe thermal prismatic block small modular reactor using TRISO high-assay low-enriched uranium (HALEU) as fuel, helium as coolant, and graphite as a moderator. Its design includes passive safety systems and robotic refueling. The first of the two FMRs is a high-temperature (850°C) gas-cooled fast reactor (GFR) module of 50 MWe using HALEU (19.75 wt.%) UO_2 pellets in silicon carbide composite cladding. The second is the Energy Multiplier (EM^2), a small module of 265 MWe using uranium carbide (UC) annular pellets in silicon carbide composite cladding.[22] EM^2 converts fertile isotopes to fissile ones and burns them in situ throughout the 30 years of core life. Both FMRs use helium to cool the core and, also, to spin the turbines and generate electricity, and therefore require no steam generator [85, 86].

There are also some other advanced technologies currently undergoing pre-application activities with the NRC. One of those is the Kairos Power fluoride-salt-cooled high-temperature reactor (KP-FHR), a small modular fast reactor of 140 MWe using TRISO pebbles floating in the FLiBe salt. This TRISO fuel is 19.75% ^{235}U enriched (HALEU). The reactor concept combines elements of high-temperature gas-cooled reactors and molten salt reactors and works at atmospheric pressure [87]. The combination of an accident-tolerant fuel like TRISO, the low-pressure design, and the passive decay heat removal and safety systems create an inherently safe design. See a schematic of this reactor concept in Figure 5.22.

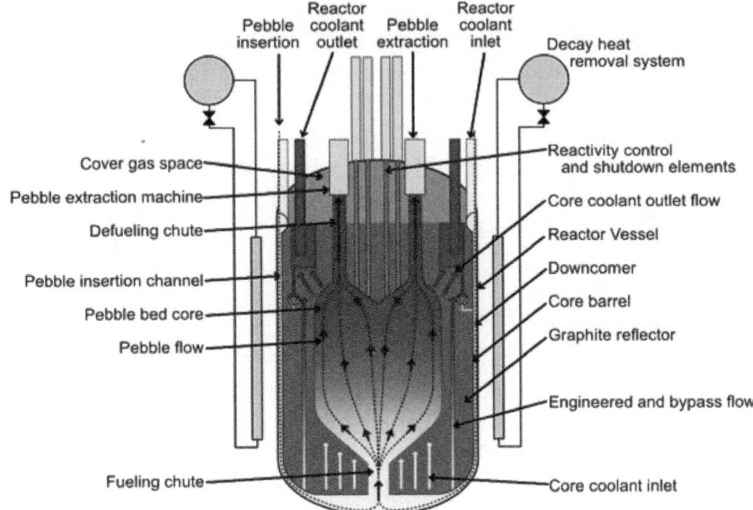

Figure 5.22. Schema of the Kairos Power FLiBe salt-cooled high-temperature reactor (FHR) (Image courtesy of the NRC).

---

[22] This new fuel type is currently in pre-application engagement with NRC for fuel qualification.

Kairos is also in the process of licensing the construction of the Hermes Test Reactor in Oak Ridge—a low-power scaled version of the KP-FHR designed to support the FLiBe salt-cooled, high-temperature reactor technology research and development. Hermes, which also uses HALEU TRISO fuel, will not produce electricity [88].

Another advanced technology that is being subject to pre-application activities with the NRC is the Natrium™ Reactor. This is a small modular sodium fast reactor of 345 MWe, pool-type, using HALEU metal fuel. The reactor is coupled with an integrated molten salt-based energy storage system that can respond to spikes and drops in electricity demand. The Natrium reactor operates at high temperatures and near atmospheric pressure and is part of TerraPower's Natrium project [89].

TerraPower's project also includes pre-application with the NRC for the Molten Chloride Fast Reactor (MCFR), a small modular fast reactor, where a mixture of sodium chloride (NaCl) and highly enriched uranium trichloride ($^{235}UCl_3$) serves both as a coolant and fuel. The technology of this uranium chloride salt-fueled concept will be demonstrated at the zero-power testbed at Idaho National Laboratory (INL) [90]. Like Natrium, the MCFR reactor is also expected to be paired with a molten salt energy storage system. Molten salt reactors can use higher temperatures, thus improving the operating efficiency. Also, some designs can breed and burn fuel reducing the volume of waste [85].

Terrestrial Energy's Integral Molten Salt Reactor® (IMSR) has completed phase two of the pre-licensing vendor design review with the Canadian Nuclear Safety Commission (CNSC) and is now undergoing pre-application activities with the NRC. It is a 195 MWe small modular thermal reactor, cooled by a fluoride molten salt mixture [lithium fluoride (LiF), sodium fluoride (NaF), and/or beryllium fluoride (BeF_2)]. The fuel is liquid uranium tetrafluoride ($^{235}UF_4$) with less than 5% enrichment, which is blended with the salt mixture.

As illustrated in Figure 5.23, the most notable feature of the IMSR is the integration of all the primary reactor components—fuel salts, the graphite moderator, pumps, and heat exchangers—into a sealed and replaceable reactor vessel called the Core-unit, that is manufactured outside the nuclear plant, in a controlled factory environment [91].

At the reactor site, the Core-unit is lowered into a guard vessel located in a below-grade reactor silo and is connected to the secondary piping also containing fluoride salt but without fuel. The guard vessel is the containment and passive cooling for the Core-unit. The service life of the IMSR® Core-unit is 7 years, after which it is replaced by a new one [92].

At the end of 2022, Abilene Christian University applied to the NRC for a license to build a molten salt research reactor (MSRR). The purpose of

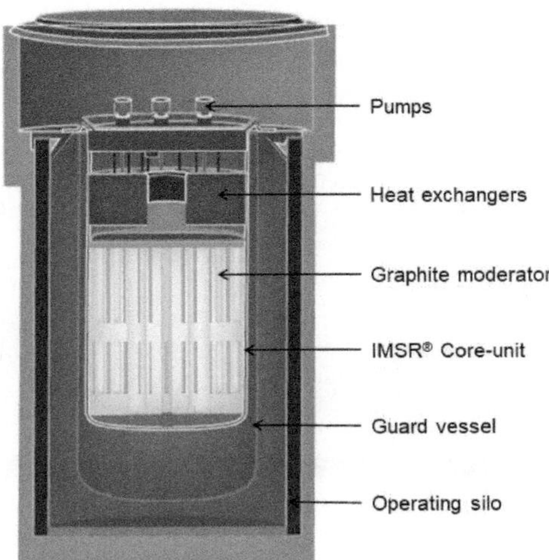

Figure 5.23. IMSR® Core-unit inside a reactor silo (Creative-common image courtesy of Terrestrial Energy).

the Abilene Christian University's Nuclear Energy eXperimental Testing (NEXT) Lab is to support academic research on thermal high-temperature molten salt reactors, using HALEU fuel dissolved in FLiBe salt. The proposed reactor will be a graphite-moderated, fluoride-salt-cooled, and liquid fuel research reactor [93].

The ARC-100 from ARC Clean Technology is another small modular reactor, but sodium-cooled. This reactor was chosen for deployment at the New Brunswick Power of Canada Point Lepreau site. ARC-100 is a 100 MWe pool-type sodium fast reactor. It uses HALEU metallic uranium alloy fuel, enriched to less than 20%. After the initial load of uranium fuel, the reactor will breed as much fissile fuel as it consumes, for as long as 20 years. ARC-100 parts can be built at a centralized factory and then assembled at the site. An advantage with this reactor is that it significantly reduces the amount of long-term waste produced [94]. Currently, the ARC-100 is in design review with the CNSC and [95] is undergoing pre-application activities with the NRC [96].

New advanced reactors include microreactors. These are a subset of small modular reactors with a power output in the range of 1 to 20 MWt (1–50 MWe). Microreactors also use different reactor technologies from heat-pipe-cooling to high-temperature-gas-cooling and fast neutrons. They can be connected to the electric grid or a microgrid with renewable sources, or even operate without support from a surrounding transmission grid. Microreactors have the potential to produce electricity and heat for remote communities, mines, heavy industries, and military bases, and even serve for disaster relief [97].

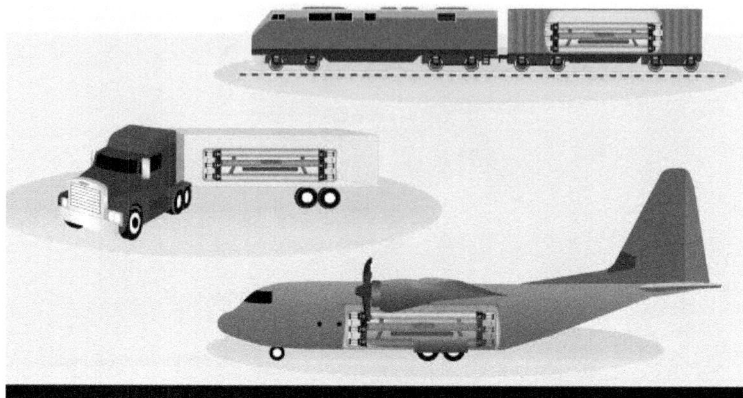

Figure 5.24. Microreactors can be easily transported by rail, truck, or cargo plane (Image courtesy of U.S. Government Accountability Office).

Some of the advantages of microreactors are their small size, being factory-built, not needing intricate construction work, being easily assembled on-site, being transportable (Figure 5.24), and having long refueling intervals or no refueling at all [98, 99].

Some microreactors that currently are in pre-conceptual design or conceptual design are Energy Well from the Czech Republic, U-Battery from the UK, MoveluX from Japan, and Elena from Russia [99]. However, only a few mentioned below have applied for authorization from the regulatory authority. These are expected to be deployed by 2030. They are the Micro Modular Reactor (MMR) from Ultra Safe Nuclear Company—a partnership between the US and Canada; the eVinci™ from Westinghouse, and the Aurora Powerhouse from Oklo Inc., in the US. The MMR™ already completed the vendor design review with the CNSC and is in the process of licensing the construction and operation of a demo microreactor at the Canadian Nuclear Laboratories (CNL) in Chalk River [100]. eVinci™ and Aurora are in pre-application review with the NRC [101].

The MMR™ is a helium-cooled, high-temperature microreactor (HTGR) of 5 MWe, using HALEU (19.7% ^{235}U) fully microencapsulated TRISO pellets as fuel, stacked in a core built of hexagonal graphite blocks. The fuel life is expected to last 20 years, although the core can then be replaced to extend the operating life of the reactor for another 20 years. The safety features of the MMR™ include both passive and active systems. For example, if a cooling failure occurs, the reactor will be cooled through passive heat transfer; and, in case of an immediate shutdown, the control rods will fall by gravity and cut the power to the helium circulator motor making the reactor subcritical [102].

In the MMR™ design, the nuclear plant and the non-nuclear power plant are separated. This means that the primary heat transfer loop and the steam generation loop are divided by way of an intermediate molten salt storage loop to produce the steam. The reactor is built below grade for external hazard protection. It also houses the equipment to transfer the heat to the adjacent non-nuclear power plant [103].

The University of Illinois also applied to the NRC for a permit to build an MMR™ as a testing facility on the Urbana-Champaign campus. The idea is to have an advanced reactor at a research, education, and training facility to do experiments, prototype tests, and simulations as a tool for workforce education and training. Furthermore, the heat stored in the intermediate loop will interface with the University's electrical and steam distribution system and produce steam and electricity for the University's own needs [104].

eVinci™ is a 5 MWe thermal microreactor using TRISO fuel in a prismatic-graphite block and an innovative fully passive, alkali metal heat pipe-cooled design to flexibly operate on the grid and in off-grid locations, such as mining sites and remote communities. The heat pipes are embedded in channels in the reactor's graphite core, alongside TRISO fuel pins, and shutdown rods. Thick radial control drums are located around the core for reactivity control and shutdown. The core and the control drums, surrounded by a neutron shield and a gamma shield, are all enclosed in a canister containment.

The passive heat pipe-cooled design consists of long, thin fully sealed tubes, that passively transfer thermal energy from one end of the tube to the other. Each heat pipe contains a small volume of liquid sodium which is first turned into vapor and then condensed back to return to the hot interface through capillary action, releasing its latent heat[23] [105]. A schematic of how heat pipes work is shown in Figure 5.25. Heat pipes are made of a particular alloy of iron, chromium, and aluminum (FeCrAl),

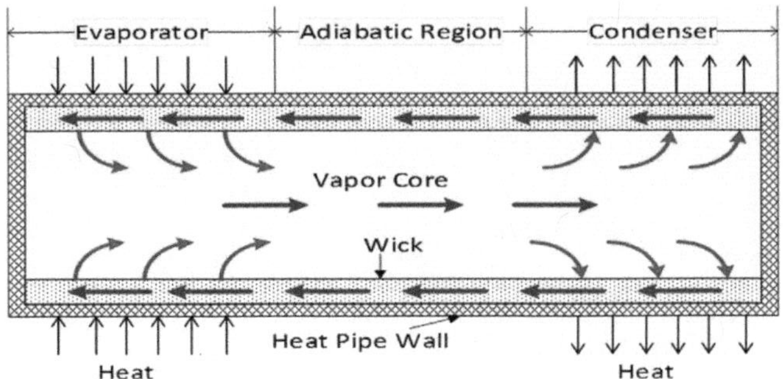

Figure 5.25. Diagram of the heat transfer in passive heat pipes.

---

[23] Latent heat is the hidden energy that changes the state of matter without changing its temperature.

and were developed and tested by Los Alamos National Laboratory for space applications.

eVinci™ is currently in pre-application review with both the NRC and the CNSC for certification of its design and technology. eVinci™ microreactor is also fully contained, factory-built, transportable, and generates electricity and heat. Designed for passive, autonomous operations, eVinci™ is expected to generate electricity 24 hours a day, 7 days a week, and 365 days a year, for eight or more years without refueling, performing like a solid-state nuclear battery [106].

The Aurora power plant is a fast reactor designed to produce up to 15 MWe based on the experience of the experimental breeder reactor from the 1960s. It uses HALEU U-10Zr or recycled fuel, i.e., it can recycle its own discharged fuel and reduce the volume of waste. A 10-year operation is expected with a single fuel load. As illustrated in Figure 5.26, the reactor is to be set inside a building known as the Aurora Powerhouse. The A-frame building with a relatively small area of about 4000 to 8000 m^2 houses the reactor in the basement. The power conversion system and other supporting equipment are planned to be located on the ground floor.

Table 5.3 below summarizes some data about large evolutionary water-cooled reactors (the so-called generation 3+) recently connected to the grid as per the reactor database of the World Nuclear Association [107]. See also Table 5.2 for reference.

As shown in Table 5.3, four advanced water-cooled reactors were connected to the grid till June 2023. The remaining six were connected in 2022 and are currently in commercial operation.

Figure 5.26. A computer-generated image of Aurora Powerhouse by Gensler (Photo reproduced with Oklo permission).

Table 5.3. New reactors in the world recently connected to the grid.

Reactor name	Generation 3+ model	Type	Net Capacity (MWe)	Connection date	Country
Ostrovets 2	VVER V-491	PWR	1110	2023-05-13	Belarus
Vogtle 3	AP1000	PWR	1117	2023-04-01	USA
Mochovce 3	VVER V-213	PWR	440	2023-01-31	Slovakia
Fangchenggang 3	HPR1000	PWR	1105	2023-01-10	China
Barakah 3	APR-1400	PWR	1337	2022-10-08	United Arab Emirates
Shin Hanul 1	APR-1400	PWR	1414	2022-06-09	South Korea
Hongyanhe 6	ACPR-1000	PWR	1061	2022-05-02	China
Olkiluoto 3	EPR	PWR	1600	2022-03-12	Finland
Karachi 3	HPR1000	PWR	1017	2022-03-04	Pakistan
Fuqing 6	HPR1000	PWR	1075	2022-01-01	China

The only liquid metal fast reactors in commercial operation thus far are the early described sodium-cooled BN-600 and BN-800 in Beloyarsk, Russia. However, various non-water reactors are in different levels of design. According to the IAEA Advanced Reactors Information System (ARIS), five thermal gas-cooled reactors (GCR) are in development in the US, Japan, China, and South Africa. The one in China is under construction, the one in South Africa is on hold, and the other three are still under design. Three fast gas-cooled reactors (GFR) are in development in Europe, the US, and Japan, and all of them are under design [77].

For example, the first two modules of the Chinese high-temperature, gas-cooled reactor HTR-PM600 reached criticality and were connected to the grid in Shandong, in 2021. The HTR-PM600 is a 600 MWe demonstration nuclear plant, based on the 100 MWe small module HTR-PM (high-temperature gas-cooled pebble-bed). To get an output of 600 MWe, the plant uses six such modules arranged in three units; each unit with two HTR-PM and a single 210 MWe steam turbine. The coolant in the HTR-PM is helium and the moderator is graphite. The fuel consists of over 245,000 pebbles made of TRISO particles, each 60 mm in diameter, and containing 7 g of $^{235}UO_2$ enriched to 8.5% [108].

Regarding molten-metal-cooled reactors, ten fast sodium-cooled reactors have been in development in Japan, France, Russia, China, India, South Korea, and the US but only one of them, the Russian BM-1200 is under construction. In addition, twelve lead-cooled reactors were planned in Europe, Russia, China, Sweden, Luxembourg, Belgium, South Korea, and the US, but all are only in design [77].

Two examples include the French sodium-cooled fast reactor ASTRID (Advanced Sodium Technological Reactor for Industrial Demonstration) and the Argonne National Laboratory lead-cooled fast reactor SSTAR (Secure Transportable Autonomous Reactor).

ASTRID conceptual design was completed in 2015. It was at first envisaged as a 600 MWe prototype to demonstrate the feasibility of the transmutation of minor actinides. However later, its capacity was reduced to 100–200 MWe. The variety of fuel it could use—spent fuel from operating commercial reactors, MOX, and depleted and natural uranium—was one of its advantages. The project was abandoned in 2019.

The Secure Transportable Autonomous Reactor (STAR) was first conceived in 1997 and later resulted in two transportable autonomous small reactors, the Secure Transportable Autonomous Reactor with Liquid Metal (STAR-LM) and the Small Secure Transportable Autonomous Reactor (SSTAR), for deployment on remote or small electrical power grids. Both STAR-LM and SSTAR were modular, factory-built fast reactors, proliferation-resistant, and suitable for developing countries. These would use nitride fuel, and natural circulation, and have a long refueling interval of 15–30 years. The vessel was sealed and designed for a complete cassette-core replacement for refueling [109]. A prototype was scheduled for manufacture in 2015 but was abandoned. Nevertheless, the SSTAR design has been retained as the Generation IV International Forum (GIF) reference for a small modular LFR system [110].

It is expected that advanced reactors will operate with significantly enhanced safety compared to traditional light-water nuclear reactors. These reactors benefit from passive safety systems functioning by natural forces, such as gravity, pressure differences, or natural heat convection. Besides, they can operate at higher temperatures and lower, safer pressures for more efficiency and safety. Advanced reactors can also use various fuels while simultaneously addressing proliferation and radioactive waste generation concerns. Furthermore, small modular advanced reactors and microreactors can be installed in underground caverns where radiation and security risks could be reduced, and they can also vary the amount of power they produce to balance electricity loads.

At present, only two SMRs are in operation in the world. They are the Russian floating nuclear cogeneration plant Akademik Lomonosov, which was connected to the grid at the end of 2019 and commissioned in 2020, and the high-temperature pebble HTR-PM module in China, whose first two units were connected to the grid at the end of 2021.

References

[1] Touran, N. 2022. What is a nuclear reactor? Whatisnuclear.com, 2022. [Online]. Available: https://whatisnuclear.com/reactors.html. [Accessed 27 January 2023].

[2] US NRC. 2022. Power Reactors. Nuclear Regulatory Commission, 16 September 2022. [Online]. Available: https://www.nrc.gov/reactors/power.html. [Accessed 24 January 2023].

[3] PRIS Power Reactor Information System. 2023. The Database on Nuclear Power Reactors. International Atomic Energy Agency, 23 January 2023. [Online]. Available: https://pris.iaea.org/pris/. [Accessed 24 January 2023].

[4] World Nuclear Association. 2017. Mixed Oxide (MOX) Fuel. World Nuclear Association, October 2017. [Online]. Available: https://www.world-nuclear.org/information-library/nuclear-fuel-cycle/fuel-recycling/mixed-oxide-fuel-mox.aspx. [Accessed 26 January 2023].

[5] World Nuclear Association. 2021. Nuclear Fuel and its Fabrication. World Nuclear Association, October 2021. [Online]. Available: https://www.world-nuclear.org/information-library/nuclear-fuel-cycle/conversion-enrichment-and-fabrication/fuel-fabrication.aspx. [Accessed 29 January 2023].

[6] The Institution of Electrical Engineers (Now IET). 1993. Nuclear Reactor Types. An Environment & Energy FactFile provided by the IEE. The Institution of Electrical Engineers, London.

[7] World Nuclear Association. 2022. RBMK Reactors – Appendix to Nuclear Power Reactors. World Nuclear Association, February 2022. [Online]. Available: https://www.world-nuclear.org/information-library/nuclear-fuel-cycle/nuclear-power-reactors/appendices/rbmk-reactors.aspx. [Accessed 1 February 2023].

[8] Westinghouse Electric Company. 2016. The Art of Innovation: Westinghouse AGR Fuel. Westinghouse, 30 June 2016. [Online]. Available: https://info.westinghousenuclear.com/blog/the-art-of-innovation-westinghouse-agr-fuel. [Accessed 1 February 2023].

[9] Nonbel, E. 1996. Description of the Advanced Gas Cooled Type of Reactor (AGR). Rise National Laboratory, Roskilde.

[10] McKeen, T. 2010. Advanced Gas Cooled Reactors. pp. 1–28. In Nuclear Energy Materials and Reactors - Volume II, Oxford, Eolss Publishers.

[11] Canadian Nuclear Safety Commission. 2022. Pressure tubes: The heart of the reactor. Canadian Nuclear Safety Commission, 10 October 2022. [Online]. Available: http://www.nuclearsafety.gc.ca/eng/resources/fact-sheets/pressure-tubes.cfm. [Accessed 31 January 2023].

[12] Garland, B. 2013. CANDU Fundamentals - Bill Garland's Nuclear Engineering Page. canteach.candu.org, 8 February 2013. [Online]. Available: https://www.yumpu.com/en/document/view/18655428/candu-fundamentals-bill-garlands-nuclear-engineering-page/126. [Accessed 1 February 2023].

[13] Bordia, P. 2012. CANDU Reactors, Submitted as coursework for PH241 Introduction to Nuclear Energy. Stanford University, 2012. [Online]. Available: http://large.stanford.edu/courses/2012/ph241/bordia2/#:~:text=CANDU%20stands%20for%20CANada%20Deuterium%20Uranium%20is%20a,a%20nuclear%20fission%20reaction%20is%20an%20~MeV%20neutron.. [Accessed 1 February 2023].

[14] Simopt LLC. 2023. Energy Encyclopedia, The Nuclear Reactors. Simopt LLC, 2023. [Online]. Available: https://www.energyencyclopedia.com/en/nuclear-energy/the-nuclear-reactors. [Accessed 2 February 2023].

[15] U.S. Department of Energy. 1993. Nuclear Physics and Reactor Theory Handbook DOE-HDBK-1019, Washington: U.S. Department of Energy.

[16] International Atomic Energy Agency. 2016. IAEA Specific Safety Requirements No. SSR-2/1 (Rev. 1), Safety of Nuclear Power Plants: Design, Vienna: IAEA.

[17] International Atomic Energy Agency. 2019. Specific Safety Guide No. SSG-53, Design of the Reactor Containment and Associated Systems for Nuclear Power Plants, Vienna: IAEA.

[18] US NRC. 2017. Reactor Concepts (R-100) Training Course, Chattanooga: NRC Technical Training Center.

[19] US NRC. 2012. SECY-12-0157 - Enclosure 2 - Boiling Water Reactor Mark I and Mark II Containment Regulatory History. US NRC, Washington.

[20] nuclear-power.com. 2023. Containment Building. nuclear-power.com, 2023. [Online]. Available: https://www.nuclear-power.com/nuclear-power-plant/containment-building/. [Accessed 9 February 2023].

[21] CANTEACH. 2011. CANDU Nuclear Training Course 233, Reactor & Auxiliaries Module 13, Containment, Toronto: CANTEACH.

[22] Joyce, M. 2018. Mainstream power reactor systems. pp. 227–261. In Nuclear Engineering: A Conceptual Introduction to Nuclear Power, Amsterdam, Elsevier.

[23] Hughes, H. and R. Hargreaves. 1985. AGR fuel pin pellet-clad interaction failure limits and activity release fractions (IWGGCR--8). IAEA, Vienna.

[24] International Atomic Energy Agency. 1992. Safety Series No. 75-INSAG-7 The Chernobyl Accident: Updating of INSAG-1 A report by the International Nuclear Safety Advisory Group, Vienna: IAEA.

[25] Westfall, C. 2004. Vision and reality: The EBR-II story. Nuclear News, pp. 25–32, Volume 47, Number 2 February 2004.

[26] Michal, R. 2001. Fifty years ago in December: Atomic reactor EBR-I produced first electricity. Nuclear News, 44: 28–29. 12 November 2001.

[27] Michal, R. 2001. Koch: Remembering the EBR-I (l. Koch interview by Rick Michal). Nuclear News, 44: 30–35. 12 November 2001.

[28] Argonne National Laboratory. 2014. Nuclear Engineering Division, Argonne's Nuclear Science and Technology Legacy. U.S. Department of Energy Office of Science, 7 February 2014. [Online]. Available: https://www.ne.anl.gov/About/hn/logos-winter02-psr.shtml. [Accessed 5 March 2023].

[29] Argonne National Laboratory. 2007. Reactors Designed by Argonne National Laboratory. US DOE Office of Science, 10 November 2007. [Online]. Available: https://www.ne.anl.gov/About/reactors/integral-fast-reactor.shtml. [Accessed 17 March 2023].

[30] World Nuclear Association. 2021. Fast Neutron Reactors. World Nuclear Association, August 2021. [Online]. Available: https://world-nuclear.org/information-library/current-and-future-generation/fast-neutron-reactors.aspx. [Accessed 25 January 2023].

[31] International Nuclear Safety Center (INSC) database. 2006. Overview of Fast Reactors in Russia and the Former Soviet Union. 28 July 2006. [Online]. Available: https://fissilematerials.org/library/insc.pdf. [Accessed 3 March 2023].

[32] World Nuclear News WNN. 2016. Russia's BN-800 unit enters commercial operation. World Nuclear Association, 1 November 2016. [Online]. Available: https://www.world-nuclear-news.org/NN-Russias-BN-800-unit-enters-commercial-operation-01111602.html. [Accessed 11 March 2023].

[33] Pakhomov, I. 2018. BN-600 and BN-800 Operating Experience. In Gen IV International Forum, https://gif.jaea.go.jp/webinar/Series24/gifiv_webinar_pakhomov_19_dec_2018_final.pdf.

[34] World Nuclear News WNN. 2022. Beloyarsk BN-800 fast reactor running on MOX. World Nuclear Association, 13 September 2022. [Online]. Available: https://www.world-nuclear-news.org/Articles/Beloyarsk-BN-800-fast-reactor-running-on-MOX. [Accessed 5 March 2023].

[35] Schneider, M. 2009. Fast Breeder Reactors in France. Science and Global Security 17(1): 36–53.

[36] International Atomic Energy Agency. 2012. Status of Fast Reactor Research and Technology Development, IAEA-TECDOC-1691, Vienna: IAEA.

[37] International Atomic Energy Agency. 1998 International Nuclear and Radiological Event Scale (INES). IAEA. [Online]. Available: https://www.iaea.org/resources/databases/international-nuclear-and-radiological-event-scale. [Accessed 16 June 2023].

[38] Jensen, S. E. and P. Elgaard. 1995. NKS/RAK-2(95)TR-C1, Description of the Prototype Fast Reactor at Dounreay. Rise National Laboratory, Roskilde.

[39] Cruickshank, A. and A. Judd. 2000. Problems Experienced During Operation of the Prototype Fast reactor, Dounreay, 1974–1994. In Unusual occurrences during LMFR operation, Proceedings of a Technical Committee meeting held in Vienna, 9-13 November 1998, Vienna.

[40] International Atomic Energy Agency. 2007. IAEA-TECDOC-1569 Liquid Metal Cooled Reactors: Experience in Design and Operation, Vienna: IAEA.

[41] International Atomic Energy Agency. 2009. INFCIRC/754, Agreement between the Government of India and the International Atomic Energy Agency for the Application of Safeguards to Civilian Nuclear Facilities, Vienna: IAEA.

[42] International Atomic Energy Agency. 2023. INFCIRC/754/Add.12, Agreement between the Government of India and the International Atomic Energy Agency for the Application of Safeguards to Civilian Nuclear Facilities, Addition to the List of Facilities Subject to Safeguards Under the Agreement, Vienna: IAEA.

[43] Vijayan, P., I. Dulera, P. Krishnani, K. Vaze, S. Basu and R. Sinha. 2013. Overview of the Thorium Programme in India. In iThEC13-International Thorium Energy Conference, Geneva.

[44] International Atomic Energy Agency. 2023. PRIS Power Reactor Information System - India. IAEA, 22 March 2023. [Online]. Available: https://pris.iaea.org/PRIS/CountryStatistics/CountryDetails.aspx?current=IN. [Accessed 23 March 2023].

[45] Dey, P. 2003. Spent Fuel Reprocessing: An Overview. In Conference of Indian Nuclear Society 2003, Kalpakkam.

[46] World Nuclear Association. 2023. Nuclear Power in India. World Nuclear Association, January 2023. [Online]. Available: https://world-nuclear.org/information-library/country-profiles/countries-g-n/india.aspx. [Accessed 23 March 2023].

[47] Nuclear Asia. 2022. India's prototype fast breeder reactor delayed further, likely to be commissioned in 2024. Nuclear Asia, 21 December 2022. [Online]. Available: https://www.nuclearasia.com/news/indias-prototype-fast-breeder-reactor-delayed-further-likely-to-be-commissioned-in-2024/4912/. [Accessed 23 March 2023].

[48] US NRC. 2023. Accident Tolerant Fuels. Nuclear Regulatory Commission, 9 February 2023. [Online]. Available: https://www.nrc.gov/reactors/power/atf.html. [Accessed 4 April 2023].

[49] Energy.gov. 2018. GE's Nuclear Fuel Designs Ready for Reactor Testing. US Department of Energy, 16 February 2018. [Online]. Available: https://www.energy.gov/ne/articles/ges-nuclear-fuel-designs-ready-reactor-testing. [Accessed 13 February 2023].

[50] Energy.gov. 2018. Industry's First Complete Accident Tolerant Fuel Assembly in Operation at Commercial U.S. Reactor. Office of Nuclear Energy, US Department of Energy, 8 November 2018. [Online]. Available: https://www.energy.gov/ne/articles/industrys-first-complete-accident-tolerant-fuel-assembly-operation-commercial-us. [Accessed 2 April 2023].

[51] Goodson, C. and K. Geelhood. 2020. PNNL-30445, Degradation and Failure Phenomena of Accident Tolerant Fuel Concepts. FeCrAl Alloy Cladding. Pacific Northwest National Laboratory, Richland.

[52] Khatib-Rahbar, M., A. Krall, Z. Yuan and M. Zavisca. 2020. ERI/NRC 20-209 Review of Accident Tolerant Fuel Concepts With Implications to Severe Accident Progression and Radiological Releases. Energy Research, Inc., Rockville.

[53] Lightbridge Corporation. 2023. Lightbridge presentation to investors June 2023. Lightbridge Corporation, Reston.

[54] Brachet, J.-C., M. Le Saux, T. Guilbert and M. Tupin. 2015. On-going studies at CEA on chromium coated zirconium based nuclear fuel claddings for enhanced accident tolerant LWRS fuel. In TopFuel 2015, Zurich.

[55] OECD Nuclear Energy Agency (NEA). 2018. Report No. 7317 State-of-the-Art Report on Light Water Reactor Accident-Tolerant Fuels. OECD, Paris.

[56] World Nuclear News WNN. 2023. Westinghouse ADOPT fuel gets NRC approval. World Nuclear Association, 14 March 2023. [Online]. Available: https://world-nuclear-news.org/Articles/Westinghouse-accident-tolerant-fuel-gets-NRC-appro.   [Accessed 22 June 2023].

[57] US NRC. 2023. Accident Tolerant Fuel Regulatory Activities. Nuclear Regulatory Commission, 9 February 2023. [Online]. Available: https://www.nrc.gov/reactors/power/atf/lead-test.html. [Accessed 25 May 2023].

[58] Idaho National Laboratory. 2023. Transient Testing. US Department of Energy, [Online]. Available: https://transient.inl.gov/SitePages/Home.aspx. [Accessed 10 April 2023].

[59] Nuclear Energy Institute (NEI). 2019. White Paper. The Economic Benefits and Challenges with Utilizing Increased Enrichment and Fuel Burnup for Light-Water Reactors. Nuclear Energy Institute, Washington.

[60] Mishchenko, Y., K. D. Johnson, D. Jädernäs and J. Wallenius. 2021. Uranium nitride advanced fuel: an evaluation of the oxidation resistance of coated and doped grains. Journal of Nuclear Materials, vol. 556, no. December 1, p. Article 153249.

[61] General Atomics. 2018. SiGA™ Composite Capabilities, San Diego: General Atomics.

[62] Ultra Safe Nuclear. 2023. TRISO/FCM® FUEL. Ultra Safe Nuclear Corporation, 2023. [Online]. Available: https://www.usnc.com/fuel/. [Accessed 16 April 2023].

[63] TRISO-X. LLC. 2022. TRISO-X. TRISO-X. LLC, 2022. [Online]. Available: https://triso-x.com/fuel-kernels. [Accessed 21 June 2023].

[64] Stoller, R. E., G. D. Samolyuk and T. M. Besmann. 2012. Publication. Modeling Deep Burn TRISO Particle Nuclear Fuel... Oak Ridge National Laboratory, n.d 2012. [Online]. Available: https://www.ornl.gov/publication/modeling-deep-burn-triso-particle-nuclear-fuel. [Accessed 24 October 2023].

[65] International Atomic Energy Agency. 2019. IAEA Nuclear Energy Series No. NW-T-1.7. Waste from Innovative Types of Reactors and Fuel Cycles. A Preliminary Study. IAEA, Vienna.

[66] Brown, N. R., R. Hernandez and A. T. Nelson. 2022. High volume packing fraction TRISO-based fuel in light water reactors. Progress in Nuclear Energy, vol. 146, no. April 22, p. Article 104151.

[67] Free Form Fibers. 2023. Energy. Free Form Fibers, LLC, 2023. [Online]. Available: https://www.fffibers.com/industries-applications/energy/. [Accessed 21 April 2023].

[68] Hiscox, B. 2018. Analysis and Optimization of a New Accident Tolerant Fuel Called Fuel-in-Fibers, Cambridge: Massachusetts Institute of Technology.

[69] Mariani, R. D., D. L. Porter, S. L. Hayes and J. R. Kennedy. 2012. Metallic Fuels: The EBR-II legacy and recent advances. In ATALANTE 2012 International Conference on Nuclear Chemistry for Sustainable Fuel Cycles, Montpellier.

[70] Gehin, J. C. and J. J. Powers. 2016. Liquid fuel molten salt reactors for thorium utilization. Nuclear Technology 194(2): 152–161.

[71] US NRC. 2023. Molten Chloride Fast Reactor (MCFR). US Nuclear Regulatory Commission, 1 February 2023. [Online]. Available: https://www.nrc.gov/reactors/new-reactors/advanced/licensing-activities/pre-application-activities/mcfr.html. [Accessed 26 April 2023].

[72] Clean Core Thorium Energy. 2023. Clean Core Thorium Energy's Advanced Nuclear Fuel Design Begins Canadian Nuclear Safety Commission Pre-Licensing Review. Clean Core Thorium Energy, 4 February 2023. [Online]. Available: https://cleancore. energy/news/clean-core-thorium-energys-advanced-nuclear-fuel-design-begins-canadian-nuclear-safety-commission-pre-licensing-review. [Accessed 22 June 2023].

[73] Statista Research Department. 2023. Number of commercial nuclear power reactors in the United States in 2022, by years of operation. Statista, 2 February 2023. [Online]. Available: https://www.statista.com/statistics/201685/number-of-us-nuclear-power-reactors-by-years-of-operation/. [Accessed 28 April 2023].

[74] GenIV International Forum www.gen-4.org. 2021. GIF 2021 Annual Report, Chapter 2. GIF Outlook and Current Initiatives. NEA, Paris.

[75] Generation IV International Forum (GIF). 2023. Technology Systems. GEN IV International Forum, 2023. [Online]. Available: https://www.gen-4.org/gif/jcms/c_9353/systems. [Accessed 1 May 2023].

[76] GenIV International Forum. 2021. GIF 2021 Annual Report, Chapter 3. System Summaries. Nuclear Energy Agency (NEA), Paris.

[77] International Atomic Energy Agency (IAEA). 2019. Advanced Reactors Information System (ARIS). IAEA, 19 July 2019. [Online]. Available: https://www.iaea.org/resources/databases/advanced-reactors-information-system-aris. [Accessed 25 April 2023].

[78] International Atomic Energy Agency. 2020. Advanced Large Water Cooled Reactors. A Supplement to IAEA Advanced Reactors Information System (ARIS) 2020 Edition. IAEA, Vienna.

[79] Xing, J., D. Song and Y. Wu. 2016. HPR1000: Advanced pressurized water reactor with active and passive safety. Engineering 2(1): 79–87.

[80] Gera, B., P. K. Sharma, R. K. Singh and K. K. Vaze. 2011. CFD Analysis of passive autocatalytic recombiner. Science and Technology of Nuclear Installations 2011. doi:10.1155/2011/862812, p. Article ID 862812.

[81] International Atomic Energy Agency. 2019. IAEA-TECDOC-1869. Status of Research and Technology Development for Supercritical Water Cooled Reactors. IAEA, Vienna.

[82] US NRC. 2023. Design Certification Application - NuScale US600. Nuclear Regulatory Commission, 17 February 2023. [Online]. Available: https://www.nrc.gov/reactors/new-reactors/smr/licensing-activities/nuscale.html. [Accessed 8 May 2023].

[83] GE Hitachi. 2023. BWRX-300 Small Modular Reactor Achieves Pre-Licensing Milestone in Canada. General Electric, 15 March 2023. [Online]. Available: https://www.ge.com/news/press-releases/ge-hitachi-bwrx-300-small-modular-reactor-achieves-pre-licensing-milestone-in-canada. [Accessed 27 June 2023].

[84] Mulder, E. J. 2021. Overview of X-Energy's 200 MWth Xe-100 Reactor, Washington, DC: National Academy of Sciences.

[85] Nuclear Innovation Alliance (NA). 2023. Advanced Nuclear Reactor Technology. A Primer. https://www.nuclearinnovationalliance.org/resources, Reston.

[86] General Atomics. 2019. Status Report – EM2. General Atomics, San Diego.

[87] US NRC. 2023. Kairos. Nuclear Regulatory Commission, 21 March 2023. [Online]. Available: https://www.nrc.gov/reactors/new-reactors/advanced/licensing-activities/pre-application-activities/kairos.html. [Accessed 24 May 2023].

[88] US NRC. 2023. Hermes – Kairos Application. Nuclear Regulatory Commission, 24 April 2023. [Online]. Available: https://www.nrc.gov/reactors/non-power/new-facility-licensing/hermes-kairos.html. [Accessed 24 May 2023].

[89] US NRC. 2023. Natrium. Nuclear Regulatory Commission, 16 May 2023. [Online]. Available: https://www.nrc.gov/reactors/new-reactors/advanced/licensing-activities/pre-application-activities/natrium.html. [Accessed 27 May 2023].

[90] Nuclear Newswire. 2023. Get to know MCRE, the fast-spectrum MSR from Southern and TerraPower. American Nuclear Society, 30 March 2023. [Online]. Available: https://www.ans.org/news/article-4873/get-to-know-mcre-the-fastspectrum-msr-from-southern-and-terrapower/. [Accessed 27 May 2023].

[91] Advanced Reactors Information System (ARIS). 2016. Status Report – IMSR-400. IAEA, Vienna.

[92] Terrestrial Energy USA, Inc. 2023. Principal Design Criteria for IMSR® Structures, Systems and Components, Topical Report, Revision A. Terrestrial Energy USA, Inc., Oakville.

[93] US NRC. 2023. MSRR – Abilene Christian University Application. Nuclear Regulatory Commission, 19 May 2023. [Online]. Available: https://www.nrc.gov/reactors/non-power/new-facility-licensing/msrr-acu.html. [Accessed 29 May 2023].

[94] Advanced Reactor Concepts (ARC). 2010. ARC-100. A Sustainable, Cost-Effective Energy Solution for the 21st Century, Chevy Chase: Advanced Reactor Concepts, LLC.

[95] ARC Clean Technology. 2022. ARC Clean Technology. ARC Clean Technology, 2022. [Online]. Available: https://www.arc-cleantech.com/. [Accessed 14 June 2023].

[96] US NRC. 2023. ARC-100. Nuclear Regulatory Commission, 24 May 2023. [Online]. Available: https://www.nrc.gov/reactors/new-reactors/advanced/licensing-activities/pre-application-activities/arc-100.html. [Accessed 31 May 2023].

[97] Black, G., D. Shropshire, K. Araújo and A. van Heek. 2023. Prospects for nuclear microreactors: a review of the technology, economics, and regulatory considerations. Nuclear Technology 209(Sup 1): S1–S20.

[98] Testoni, R., A. Bersano and S. Segantin. 2021. Review of nuclear microreactors: status, potentialities and challenges. Progress in Nuclear Energy 138: Article 103822.

[99] SUBKI, H. 2021. Advances in SMR and microreactors technology development. In TM on the Status, Design Features, Technology Challenges and Deployment Models of Microreactors, Virtual Event.

[100] Nuclear Newswire. 2023. Canada's first microreactor headed to a Chalk River labs parking lot. American Nuclear Society, 18 May 2023. [Online]. Available: https://www.ans.org/news/article-5019/canadas-first-microreactor-headed-to-a-chalk-river-labs-parking-lot/. [Accessed 28 June 2023].

[101] US NRC. 2023. Pre-Application Activities for Advanced Reactors. Nuclear Regulatory Commission, April and May 2023. [Online]. Available: https://www.nrc.gov/reactors/new-reactors/advanced/licensing-activities/pre-application-activities.html. [Accessed 2 June 2023].

[102] Ultra Safe Nuclear Corporation. 2021. USNC Micro Modular Reactor (MMR™ BLOCK 1) Technical Information, Seattle: Ultra Safe Nuclear Corporation.

[103] International Atomic Energy Agency (IAEA). 2022. Advances in Small Modular Reactor Technology Developments. A Supplement to IAEA Advanced Reactors Information System (ARIS) 2022 Edition. IAEA, Vienna.

[104] American Nuclear Society. 2022. New reactor on campus? UIUC's choice for research, education, and training. Nuclear Newswire, pp. 24–31, 8 April 2022.

[105] Swartz, M. M., W. A. Byers, J. Lojek and R. Blunt. 2021. Westinghouse eVinci™ Heat Pipe Micro Reactor Technology Development. In 28th International Conference on Nuclear Engineering (ICONE), Online.

[106] Arafat, Y. and J. Van Wyk. 2019. eVinci™ Micro Reactor. Our next disruptive technology. Nuclear Plant Journal 37(2): 34–37.

[107] World Nuclear Association. 2023. Reactor Database. World Nuclear Association, 2023. [Online]. Available: https://www.world-nuclear.org/information-library/facts-and-figures/reactor-database.aspx. [Accessed 8 June 2023].

[108] Zhang, Z.-Y., Y.-J. Dong, Q. Shi, F. Li and H.-T. Wang. 2022. 600-MWe high-temperature gas-cooled reactor nuclear power plant HTR-PM600. Nuclear Science and Techniques 33(8): Article 101.

[109] ANL Nuclear Engineering Division. 2016. Advanced Reactor Development and Technology. Heavy Liquid Metal Reactor Development. Argonne National Laboratory, 21 October 2016. [Online]. Available: https://www.ne.anl.gov/research/ardt/hlmr/. [Accessed 13 June 2023].

[110] Smith, C. F. and L. Cinotti. 2023. Chapter 6. Lead-cooled Fast Reactors (LFRs). pp. 195–230. In Handbook of Generation IV Nuclear Reactors (Second Edition), Cambridge, Woodhead Publishing, Elsevier.

[111] International Atomic Energy Agency. 2012. IAEA-TECDOC-1691, Status of Fast Reactor Research and Technology Development, Vienna: IAEA.

[112] US NRC. 2023. Oklo Aurora Powerhouse. Nuclear Regulatory Commission, 18 May 2023. [Online]. Available: https://www.nrc.gov/reactors/new-reactors/advanced/ licensing-activities/pre-application-activities/okla-aurora-powerhouse.html. [Accessed 1 June 2023].

[113] BWX Technologies, Inc. 2023. BWXT TRISO FUEL. BWX Technologies. Inc. 2023. [Online]. Available: https://www.bwxt.com/what-we-do/strategic-nuclear-materials/ triso-fuel. [Accessed 18 April 2023].

Chapter 6

Nuclear Reactor Safety

Nuclear reactor safety, while focusing on ensuring the protection of workers and the public against exposure to ionizing radiation[1] or radioactive material, is also concerned with the safety and reliability of the reactor systems and components. This involves the management of radiation risks under proper operating conditions, the prevention of accidents or incidents that could lead to the loss of control over a nuclear reactor core or a nuclear chain reaction, and the mitigation of their consequences if they occur [1].

Protection and safety should be achieved in nuclear reactors by design, hence the siting, design, construction, operation, and decommissioning of nuclear reactors must adhere to strict requirements and a rigorous authorization process [2]. Such requirements also encompass controversial areas such as the management of radioactive waste and the transport of radioactive material [3].

Three areas of safety measures: radiation safety, nuclear safety, and security

To protect people and the environment from the harmful effects of ionizing radiation, it is crucial to limit and control the radiation exposure of individuals and the release of radioactive material into the environment [4]. This control is achieved by constraining the likelihood of events leading to such release or exposure and mitigating the consequences if these events do occur. Three areas of safety measures are essential to maintain the corresponding levels of safety. These include radiation safety, nuclear safety, and security.

---

[1] Ionizing radiation is the radiation emitted by unstable nuclei, e.g., fission products, or natural radioactive materials; or by X-ray machines; or other radiation generators.

Radiation safety is primarily concerned with the protection of workers and the public, including the protection of the environment. Any nuclear facility should have a radiation protection program commensurate with the likelihood and magnitude of exposures. This program includes a management structure capable of responding to all aspects of radiation safety in normal situations and emergencies. The program is also a tool for the optimization of radiation protection. It helps to evaluate whether or not the efforts for protection are reasonable and adequate, and whether the doses are as low as reasonably achievable (ALARA principle).

The radiation protection program is designed based on a realistic assessment of all sources of potential exposures, both routine and reasonably foreseeable. The estimates of relevant doses and probabilities are evaluated considering shielding, containments, interlocks, ease of decontamination, ventilation, and the like. The radiation protection program also contemplates other non-radiation health and safety issues resulting from industrial hygiene, industrial safety, and fire safety.

It usually defines, among others, the following [5]:

- the duties and responsibilities of the Radiation Safety Officer,
- the corresponding classification of areas within the reactor buildings, including the required barriers and access controls,
- the dose constraints,[2] reference levels,[3] and investigation levels[4] for each identified controlled or supervised area,
- the local operating instructions, including the provisions for the reception, storage, inventory, and disposal of radioactive materials— nuclear fuel and waste,
- the control of effluents from water treatment and air filtration, and their levels for release or disposal,
- the instruments, procedures, and schedules for workplace monitoring,
- the specifications and calibration schedules for measuring instruments,

---

[2] A value of the individual dose that serves as a boundary in defining the range of options in optimization. It is usually set up as a fraction of the dose limit, based on good practice, and on what can reasonably be achieved.

[3] A level of dose, risk, or activity concentration above which it is not appropriate to plan to allow exposures to occur in an emergency or an existing exposure situation. It also depends on the prevailing circumstances, including the net benefit of avoiding preventive or protective actions that would be detrimental to living conditions.

[4] The value of a quantity such as effective dose, intake or contamination per unit area or volume, at or above which an investigation would be conducted.

- the individual dose monitoring and exposure assessment of workers, including their medical surveillance,
- the education and training programs for all the staff,
- the recording of all relevant information related to the protection and safety of the workplace, the workers, and the environment,
- the procedures to periodically audit and review the radiation program performance,
- the content, procedures, and performance of quality assurance programs, including compliance, and
- the procedures to activate the emergency response inside and outside the facility and the resources and actions planned for emergencies.

Nuclear safety is concerned with reaching proper operating conditions, the prevention of accidents, or the mitigation of their consequences. This means that, while aiming at the protection of people and the environment against the harmful effects of ionizing radiation, nuclear safety undertakes three unique responsibilities: the large quantity of radioactive material present in the reactor core that should remain confined; the significant amount of heat remaining after the reactor shutdown that should be cooled down; and the risk of a power excursion caused by super-criticality that should be avoided [6].

These are some nuclear safety requirements that should be accomplished throughout the lifetime of the reactor facility [7]:

- the review of site-related factors during siting authorization,
- the assessment and approval of the design,
- the presence of accepted systems, structures, and components important to safety,
- the performance of tests and verifications during construction and commissioning,
- the implementation of process controls acting during operation,
- the satisfactory enforcement of compliance with operating procedures,
- the continuing monitoring and inspection of the workplace and the environment, and
- limiting of foreseeable radiation exposures during shutdown and decommissioning.

Nuclear safety is likewise a broad concept that encompasses other related areas such as criticality safety, risk assessment, quality assurance, natural phenomena protection, fire protection, and emergency response.

Criticality safety is focused on the prevention of an inadvertent self-sustained nuclear chain reaction. A criticality accident could be severe and those nearby could receive lethal doses of neutron and gamma radiation in a very short time [8]. Criticality safety is achieved by reactor physics, for example, deciding the mass, volume, and enrichment of the reactor fuel; establishing the reactor fuel assembly geometry and shape; using moderators and reflectors to control the neutron flux concentration and density; and the rest of the interdependent parameters of the core [9].

Criticality is the state in which a nuclear reactor is stable, i.e., the chain reaction is self-sustained, and the reactivity is zero [10]. Accordingly, reactivity is a measure of deviation from criticality. A positive reactivity means a deviation toward super-criticality (a power increase) while a negative reactivity indicates a deviation to sub-criticality (a power decrease) [11]. Reactivity control in the reactor is achieved by using neutron poisons—materials that absorb neutrons—inserted within the fuel elements. Also, using materials called neutron reflectors, at the edge of the reactor core, to scatter back those neutrons that, else, would escape. Chemical means are used too, for example, boric acid is dissolved in the coolant water to slow or stop the chain reaction.

Risk assessment is a tool for evaluating the likelihood and severity of accidents that could affect the safety of a nuclear reactor. It uses both deterministic and probabilistic methods to identify the potential accidents and the extent of their consequences, as well as identify safety vulnerabilities [12, 13]. Probabilistic safety analyses allow the combining of multiple events and sequences to determine what can go wrong and what changes can be made to prevent it.

Deterministic and probabilistic risk assessments play a crucial role in analyzing the effects of internal and external hazards. These assessments ensure that safety levels are not compromised. Accordingly, the design considers methods for detecting these hazards and minimizing their impact. Internal hazards can include, but are not limited to [14]:

- component failures (e.g., a piping leak),
- human errors (e.g., leaving a valve open by mistake),
- accidental fires or explosions,
- missiles internally generated by falling objects,
- pipe ruptures and resulting pipe whips or high-pressure jets,
- collapse of structures or drop of heavy objects.

External hazards require an assessment of the seismic characteristics of the nuclear reactor site [15]. Other natural phenomena such as earthquakes, storms, tornadoes, tides, flooding, and tsunamis also need to be considered [16]. Accidental or intentional human-induced events like an aircraft impact, explosion, or fire must also be evaluated. The events considered depend on the specifics of the reactor's site.

There are validated computer codes to model diverse incidents and evaluate their consequences, for example, SAPHIRE (Systems Analysis Program for Hands-on Integrated Reliability Evaluation) is a code for performing probabilistic risk assessments of nuclear power plants; MELCOR (Melting Core), developed by Sandia National Laboratories, is an integral code to model the progression of severe accidents in light-water-cooled reactors, and TRACE (TRAC/RELAP Advanced Computational Engine) can analyze large/small break LOCAs[5] and system transients in both PWRs and BWRs [17].

The use of technologies and materials of high quality is also a nuclear safety requirement. Therefore, a reactor quality assurance program should introduce confidence that materials, structures, systems, and components meet the requirements, with no defects or errors, and will perform satisfactorily. This program applies to all activities affecting the reactor safety-related functions [18]. Quality assurance comprises quality control, that is, measuring, inspecting, testing, recording, witnessing, and sampling. These activities provide objective confirmation of achieving defect-free and error-free products and services.

The American Society of Mechanical Engineers (ASME) has developed a nuclear quality assurance standard named NQA-1 specifying 18 requirements, ranging from organization to auditing, for the development and implementation of a quality assurance program. This NQA-1 Standard covers the siting, design, construction, operation, and decommissioning of nuclear reactors, including nuclear power plants, small modular reactors (SMR), and advanced reactors [19].

Although there are only subtle distinctions between safety and security, it is easy to understand that security approaches are intended to prevent unauthorized access to, and actions against, nuclear and other radioactive materials, and associated facilities. Security is, therefore, concerned with malicious activities, for instance, threats, theft, illegal transfer, sabotage, and so on [20]. Security in nuclear reactors is also associated with safeguards, physical protection, cybersecurity, information security, and more [21].

---

[5] Loss of coolant accident (LOCA) is a potential accident that results in a loss of reactor coolant at a rate greater than the capability of the makeup water system to add it back in the reactor.

Safeguards, as discussed before in Chapter 2, are a group of technical measures for accountability, sealing, surveillance, reporting, and inspection intended to ensure that nuclear materials are not diverted for military purposes. For non-nuclear countries, these measures are part of the whole system for the control and accountability of nuclear materials to fulfill their responsibilities under the NPT agreements with the IAEA.

Physical protection includes physical barriers such as walls, fences, and gates; restricted access to identified locations; employee background checks; categories of identification badging; illuminated detection zones; physical patrols using highly trained, and well-armed security officers; electronic surveillance; and so forth [22].

Regarding cybersecurity, the first line of defense is isolation, hence, critical safety and security systems at nuclear energy plants are not directly or indirectly connected to the Internet. Cybersecurity also includes a strategy consisting of a defensive architecture, that is, defensive levels separated by firewalls and diodes, and a set of strict controls including restricting the use of portable media and equipment [23].

Information security is aimed at protecting classified and safeguarded information, as well as sensitive unclassified non-safeguards information. These include among others, restricted data that could assist in the design, manufacturing, or utilization of nuclear weapons; information on the physical protection of operating nuclear reactors, spent fuel shipments, radioactive material quantities, locations, and the like.

Safety systems, structures, and components

To achieve a safe operation, by design, a nuclear reactor incorporates systems, structures, and components whose tasks are controlling the reactor core, preventing accidents, and mitigating their consequences. These are termed important to safety and are associated with critical functions such as reactivity control, heat removal, and containment of radioactive materials [24]. Thus, safety systems, structures, and components important to safety should provide reasonable assurance that the reactor can be operated without undue risk to the health and safety of the public [25].

Safety systems are, for example, the control rod drive system, the coolant system, and the emergency cooling system. These encompass protection systems, safety actuation systems, and safety system support features. Protection systems monitor operations and, on sensing an abnormal condition, initiate actions to prevent damage to the reactor and the release of radioactive material. These accordingly include sensors, circuits, and actuation devices. Safety actuation systems accomplish the action requested by the protection system. For example, insert a control rod, start the safety injection pumps, etc. Safety system support features include equipment that provides services such as cooling, steam, water, electricity supply, etc., required by the protection and safety actuation systems.

Safety structures include the physical barriers that stand between the radioactive materials in the reactor core and the environment. These barriers are considered layers of containment and typically include the fuel pellets, the fuel cladding, the reactor coolant pressure boundary, and the containment (see Figure 6.1). The reactor coolant pressure boundary comprises the pressure-containing components of BWRs and PWRs, such as pressure vessels, piping, pumps, fittings, valves, and so forth [26].

Based on deterministic and probabilistic risk assessments, safety systems, structures, and components are classified according to the importance of the function they perform in each plant state, and the consequences if such function is not accomplished [27]. Plant states are defined as:

- operational states and anticipated operational occurrences (AOOs),
- postulated design-basis accident (DBA) conditions,
- beyond design-basis accident (BDBA) conditions, both without significant fuel degradation and with core melting.

The IAEA recommended classification of systems, structures, and components can be summarized as follows [27]:

Safety class	Consequences of failure of the function of systems, structures, and components	Severity of consequences
1	release of radioactive material > DBA limit exceed acceptance criteria for DBA	High
2	release of radioactive material > AOO limit exceed the design limits for AOO	Medium
3	doses to workers > authorized limits	Low

Operational states are normal power operation and operation transients at startup, depending on the reactor, hot and cold shutdown, and refueling. An anticipated operational occurrence (AOO) is an incident of moderate frequency, i.e., an event that is expected to take place one or more times during the lifetime of the reactor. An inadvertent moderator cooldown, a loss of all offsite power, or a control rod drop are AOO examples [28].

As per the NRC definition [29], a design-basis accident (DBA) is a postulated accident that, according to its design and construction, a nuclear reactor must withstand without damaging any systems, structures, and components essential for ensuring public health and safety. This type of accident is not expected to occur during the lifetime of the reactor, but because it is predictable and preventable, it should be considered in the design of the systems, structures, and components. A loss of coolant accident (LOCA) due to a major rupture of a coolant pipe; an ejection of a control rod assembly; and a steam generator tube rupture accident (SGTR) are examples of DBAs.

Beyond-design basis accidents[6] (BDBAs) are those with low probability but very high impact. These accidents are unlikely to happen, but if they do, the consequences might be severe. Two types of BDBAs can be identified: without significant fuel degradation and with core melting. The Three Mile Island in 1979, the Chornobyl accident in 1986, and the Fukushima Daiichi accident in 2011 were all beyond-design-basis accidents with core melting. Some hypothetical accidents, e.g., a station blackout event that lasts longer than the design basis but can be recovered before the core is uncovered; and a large LOCA exceeding the design-basis break size, but which can be mitigated with the emergency core cooling system could be examples of BDBAs without significant fuel degradation.

AOOs have a frequency of occurrence greater than 10^{-2}/ry[7] while DBAs have a frequency of occurrence between 10^{-2} and 10^{-6}/ry. BDBAs have a frequency of occurrence of 10^{-4} to 10^{-6}/ry for accidents without significant fuel degradation and less than 10^{-6}/ry for those with core melting [30].

Defense-in-depth safety concept

Defense-in-depth is a philosophy by which a set of multiple independent and redundant layers of defense are applied to the design, construction, and operation of nuclear reactors to compensate for potential human and mechanical failures so that no single layer, no matter how robust, is exclusively relied upon [31]. This approach is an effective way to address uncertainties in the design, construction, maintenance, and operation of equipment and to account for human error, to achieve optimal safety and the protection of workers, the public, and the environment.

Successive measures for the prevention, detection, control, and mitigation of accidents are used to this end. These include multiple safety systems that shut down the reactor, remove heat from the core, and isolate

---

[6] The IAEA refers to them as design extension conditions (DECs).
[7] ry = reactor year.

and cool the containment to prevent or limit the release of radioactive materials. These also include operating and maintenance regimes and procedures, as well as emergency management, preparedness, and response capabilities to deal with accidents and protect people and the environment [32].

As shown in Figure 6.1, five levels of defense-in-depth have been identified as linked to the accident conditions discussed earlier [33]. They are:

Level 1	to prevent deviations from normal operation and the failure of systems, structures, and components important to safety.
Level 2	to detect and control deviations from normal operational conditions to prevent anticipated operational occurrences from escalating to accident conditions.
Level 3	to prevent damage to the reactor core, or radioactive releases, and return the reactor to a safe state in postulated accident conditions.
Level 4	to mitigate the consequences of failure of the third level of defense-in-depth.
Level 5	to mitigate the consequences of radioactive releases that could potentially result from accidents.

The five levels of defense-in-depth are represented in Figure 6.1 with dashed lines surrounding the reactor. The solid lines in the diagram indicate the three levels of confinement, i.e., the barriers to the release of fission products.

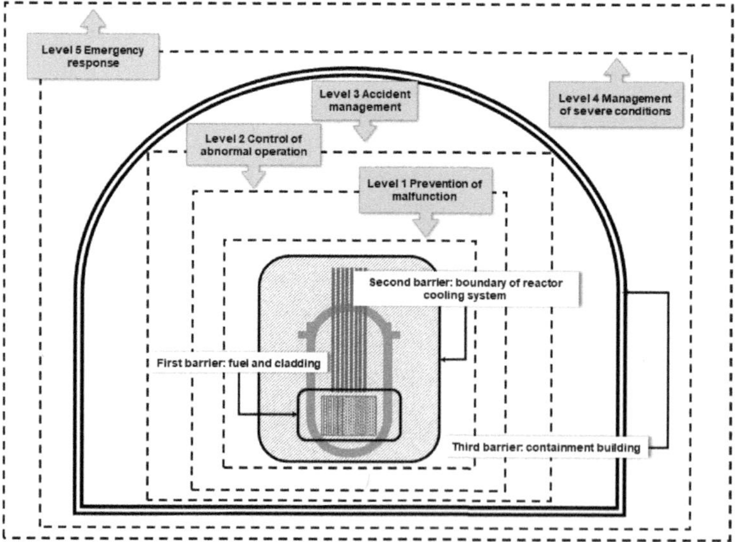

Figure 6.1. The five levels of defense-in-depth philosophy and the confinement barriers to prevent the release of fission products.

Defense-in-depth implies the existence of safety systems, structures, and components to separately protect and control the three barriers to potential releases of fission products. That is, the fuel (pellets and cladding), the reactor coolant pressure boundary (the reactor vessel, and associated piping that circulates the reactor coolant), and the containment.

In this scheme, each level should prevent degradation of the next level and mitigate the consequences of failure of the previous level. The first level prevents the malfunction of items important to safety, and the second level checks for signs of abnormal operation—anticipated operational occurrences—to restrict its progression. The third level requires actions to prevent reactor damage or radioactive releases that would call for off-site protective actions. The fourth level follows the internal emergency plan in case of the third level's failure, and, when all levels fail, the fifth level responds to the consequences of significant off-site releases as per the external emergency plan involving public authorities.

The effectiveness of a defense-in-depth strategy depends on the quality, redundancy, diversity, and independence of the systems, structures, and components important to safety and, also, the reliability of the confinement function. Safety culture is crucial in the defense-in-depth strategy to strengthen the decision-making process during normal and abnormal conditions. Lastly, emergency preparedness and response play a significant role in minimizing or eliminating long-term risks to the public and the environment.

Redundancy, diversity, and independence principles

As previously stated, the systems, structures, and components important to safety are designed to perform safety functions aimed at (1) preventing transients and accidents, (2) preventing the release of radioactive material, and (3) mitigating the consequences of accidents. Therefore, the principles of redundancy, diversity, and independence are critical to creating the successive layers of protection required to support the defense-in-depth philosophy in nuclear safety. These principles help to ensure that safety systems remain functional compensating for uncertainties and unexpected events.

Redundancy means that any system, structure, and component should have a system, structure, and component, identical or different, that will perform the same safety function no matter the state of operation or failure of the other [20]. For example, the pumps cooling the reactor core are electrically powered, but if off-site power is lost, there are emergency pumps connected to onsite diesel generators and, also, pumps connected to batteries.

Diversity means that two or more redundant systems or components should have different attributes—different physical methods, working principles, operating conditions, etc.—to reduce the chance of a common-cause failure [20]. A classic example can be found in various reactor trip capabilities: automatic control rods, manual, gravity-driven shutdown rods, and boron injection.

Independence aims at minimizing common-cause failures and the propagation of errors. It is the ability to perform the required function without being affected by the operation or failure of a redundant or diverse system or component in a different defense-in-depth level, nor by the actions to mitigate the occurrence of the postulated initiated event for which the function is required [34]. Independence is frequently combined with physical separation and electrical isolation to prevent interference.

Containment safety concept

Containment is required to fulfill the confinement of radioactive materials in operational states and accident conditions to ensure that any radioactive release to the environment is [35]:

- as low as reasonably achievable,
- below authorized limits on discharges in operational states, and
- below acceptable limits in accident conditions.

Containment encompasses the reactor or containment building which is a steel-lined reinforced concrete domed or cylindrical enclosure around the reactor. It also encompasses the systems for containment isolation, cooling, and control of combustible gases, and molten core ejection within it.

As a massive, reinforced structure, the containment building contributes to shielding against neutron and gamma radiation outside the reactor in operational and accidental conditions. Typically, there is a primary shield of water and steel around the vessel to reduce the risk of material damage and prevent the activation of components, and also to protect personnel. There is also a secondary shield surrounding the reactor coolant loops and the reactor's primary shielding [36].

Moreover, the containment building plays a role in protecting the reactor against external natural and human-induced events. As mentioned before, besides aircraft crashes, industrial explosions near the site, external missiles, and so on, these include earthquakes, floods, extreme winds, precipitations, snow, icing, lighting, fires from natural origin, and the like.

Emergency planning and response

To date, commercial nuclear reactors have been in operation for over 19,500 reactor years in 32 countries [37] and there have been three major accidents: the Three Mile Island (1979), Chornobyl (1986), and Fukushima Daiichi (2011). In the last two tragic scenarios, a large fraction of the more significant radionuclides in the respective reactors was released into the atmosphere, resulting in widespread contamination in areas of Belarus, Ukraine, and Russia in the first case, whereas in the second, it was dispersed and largely deposited over the Pacific Ocean.

In line with the International Nuclear Event Scale, both Chornobyl and Fukushima Daiichi were class 7 major accidents. The Three Mile Island accident was class 5 with wider consequences [38]. In Chornobyl, about 250,000 liquidators took part in major mitigation activities, and around 21,000 in Fukushima. These activities were at the reactor and within the zone surrounding the reactor. Of 134 individuals who suffered from acute radiation syndrome in Chornobyl, only 28 died soon after the accident from causes attributable to radiation [39].

As can be seen in Figure 6.2, the estimate of the number of deaths per TW of electricity produced by nuclear energy overall is 0.03/TWh, comparable only to modern renewable wind and solar energy. In contrast, fossil fuels and biomass account for ≈ 33% of deaths per unit of electricity. These are related to air pollution from burning fuels and the mining, transportation, and maintenance over the lifetime of the plant. Oil and gas are also worse than nuclear and renewables but to a lesser extent [40].

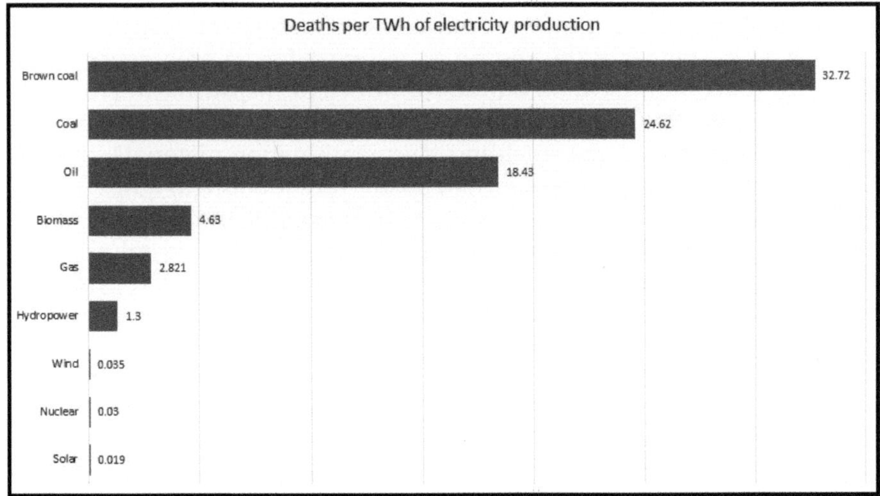

Figure 6.2. Data on death rates of electricity production from different sources (Graphic and data courtesy of Our World in Data, Creative Commons license).

From the experience of the mentioned nuclear accidents, it is clear that even the worst accident in a conventional nuclear power plant—a core meltdown—would not be likely to cause extreme public harm like, for example, the accident of a pesticide plant in Bhopal, India, in 1984, which immediate death toll was around 3000 persons. The toxic methyl isocyanate gas also caused respiratory problems, eye irritation or blindness, and other illnesses to some half a million survivors of the disaster [41].

Although the theoretical probability of accidents in nuclear reactors is around 10^{-6} ry [42], it is still a requirement for these facilities to have in place on-site and off-site emergency plans to protect workers and the public. These plans are continually updated based on the results of computer simulations, and functional exercises.

The first step of emergency planning is to define the possible scenarios, that is, the potential consequences of postulated events. These criteria, first considered in the reactor design, will guide the levels of alert and emergency, including when and where to activate the emergency control center, the alternative reactor control room if necessary, and the evacuation of all non-essential personnel from the site.

The emergency plan also clearly identifies the persons responsible for each decision, and those who will be in charge of public communications. A detailed account of the roles and responsibilities is then provided in the plan, including the names of the individuals with adequate authority to classify any incident and, upon classification, promptly, and without consultation, initiate the response. The plan provisions also specify the deadlines for notification to the off-site officials and first responders, the means of communication, the information to provide, the staff that will handle the emergency, and the resources, and materials to support them.

It is important that the resources assigned for an emergency at the reactor site—communication, and alerting devices, measuring instruments, personal protective equipment (PPE), safety signs and labels, remote recovery tools, additional shielding, decontamination materials, first aid kits, empty containers for waste, and so on—are identified, quick located and available to all concerned parts, and in good standing at all times. Since local, state, and national authorities require updated information, the appropriate communications arrangements are attached to and form part of the emergency plan. All arrangements with local authorities—police, firefighters, hospitals, etc.—are also part of the plan [5].

The off-site emergency response generally includes three possible actions involving the population near the reactor site: sheltering; thyroid chemical protection with potassium iodide; and, if it is justified by the averted dose,[8] evacuation of the population living in zones that could be affected by the radioactive release or the levels of external radiation.

---

[8] The reduction in radiation dose to the public by implementation of protective actions.

In the US, the NRC has established emergency classifications according to potential or actual effects or consequences. These increase in severity from NOUE (notification of unusual event) without radioactive release; ALERT when releases are expected to be minimal and far below the limits established by the Environmental Protection Agency (EPA); SAE (site area emergency), where major failures of plant functions, intentional damage, or malicious acts impair either equipment or, public protection; and GE (general emergency), when the event has caused substantial reactor core damage, with the potential for uncontrolled releases of radioactive material that exceeds EPA exposure levels beyond the site boundary [43]. The NRC also provides expert consultation, support, and assistance. If the incident (accident) could affect the general public, the Department of Homeland Security's Federal Emergency Management Agency (FEMA) may assume coordination of the federal response [44].

To facilitate prompt assistance and support in case of a nuclear or radiological accident at the request of any country, and in agreement with the Conventions on Early Notification of a Nuclear Accident [45] and Assistance in the Case of a Nuclear Accident or Radiological Emergency [46], the IAEA created an Incident and Emergency Center (IEC) in Vienna, to provide round-the-clock help in dealing with nuclear and radiological events, regardless of whether they arise from accident, negligence, or deliberate act. The IEC, as the headquarters for international emergency preparedness and response, coordinates the efforts, contributions, and actions of experts within the IAEA, Member States, and international organizations such as the World Health Organization (WHO), the World Meteorological Organization (WMO), and the Food and Agriculture Organization (FAO) [47, 48].

Training plays a significant role in emergency response preparedness. It helps to follow emergency procedures, handle unexpected situations, and improve confidence. Training is conducted via full-scale and functional exercises, drills, tabletop exercises,[9] seminars, workshops, and courses, on potential emergencies.

Typically, emergency training of reactor staff in specific procedures is recommended annually. For example, radiation measuring, decontamination, use of PPE, and the like. Specific training should also be provided to the local personnel that could be involved in the emergency response, for example, police, security forces and firefighters, clinicians, health practitioners and hospital staff, volunteer organizations, etc.

Full-scale and functional exercises are major-scale accident simulations, which are used to allow different groups and organizations

---

[9] An informal, discussion-based session, in which a team discusses their roles and responses during an emergency, walking through one or more example scenarios.

to act and interact in coordination [49]. NRC requires full-scale exercises at commercial nuclear reactors to be performed at least once every two years [50]. These exercises are evaluated by NRC inspectors and the Federal Emergency Management Agency (FEMA) evaluators [51].

Safety culture

Although nuclear reactors are designed, built, and operated in a safe, dependable, and effective manner, adopting a safety culture is still crucial to account for the human factor in their safety. Safety culture is a conscious way of acting that involves every employee in the nuclear organization—from the board of directors to the individual contributor—to make safety their overriding priority. It is important to recall that safety means nuclear safety, radiation safety, and industrial safety.

There are eight principles for a strong safety culture [52]:

- Everyone is personally responsible for safety.
- Leaders demonstrate commitment to safety.
- Trust permeates the organization.
- Decision-making reflects safety first.
- Nuclear technology is recognized as special and unique.
- A questioning attitude is cultivated.
- Organizational learning is embraced.
- Nuclear safety undergoes constant examination.

These principles are aimed at defining the authority and responsibility for safety at all levels of the organization. Moreover, they express their commitment to safety and to minimizing harm, clearly stating, that, from the top executives to the last employee, no one is exempt from the obligation to ensure safety first. Leaders should demonstrate their commitment to safety both in word and action when assigning priorities, and resources, and evaluating results. Managers and supervisors should be personally responsible for overseeing that all safety procedures and measures are applied and consistently reinforce the expected worker behavior. Employees are welcome to propose innovative solutions and be alert to conditions or actions that can have an undesirable impact on safety.

All the above may sound elusive, but it is not when a high level of trust is established in the organization, i.e., safety issues are freely raised and addressed timely and accurately by employees and management alike. This means that the leaders make decisions supporting a safe, reliable plant operation, and the operators are vested with the authority and responsibility necessary to place the plant in a safe condition when faced with unexpected or uncertain conditions. Furthermore, to improve future decisions, operational decisions should be questioned and reassessed when studying anomalies and potential adverse consequences.

Regarding the unique characteristics of nuclear technology, safety culture places special attention on measures and components of defense-in-depth and the maintenance and surveillance of reactivity control, continuity of core cooling, and integrity of fission product barriers and as essential functions and points of the reactor. Probabilistic risk analyses are considered in daily plant activities and plant change processes, and safety is constantly examined.

It is important to point out that reactor support groups such as human resources, labor relations, and business and financial planning should also understand their role as contributors to the prevailing focus on safety.

Safety in advanced reactors

As previously discussed, three fundamental safety functions are required from the reactor in any condition: stop the chain reaction, maintain the fuel cooling, and prevent the release of radioactive materials. Inherent and engineered safety features are designed for these purposes. Inherent safety features are based only on physical or chemical principles and make a certain reactor design safe against a particular accident. Examples of inherent safety features are the negative temperature coefficient in PWRs and the negative void coefficient in BWRs which are conditions to avoid reactor excursions [53]. Traditional reactors' engineered safety features, with only a few exceptions, are active, meaning they need an electrical or mechanical operation to work [42].

Since the Fukushima Daiichi accident, and also considering the lessons learned from the other two major accidents, the nuclear industry has been seeking answers to increase even further the reliability of reactor safety systems. Most proposed enhancements are associated with new advanced reactors. These have moved toward smaller, modular, factory-made reactors, with safety features that offer better functionality, flexibility, and resilience [54].

Advanced reactors are designed with more inherent and passive safety functions. For example, the modular high-temperature gas-cooled reactor (MHTGR) has stronger negative core temperature reactivity so, a transitory increase in temperature, will shut down the reactor. The majority of SMR designs rely on safety systems operating on natural forces, e.g., gravity, natural circulation, convection, and so forth. Hence, these could be fully capable of keeping the core cooled without pumps using external power (AC/DC) and may not need control rod insertion to shut down the reactor. Moreover, since they do not require off-site power for safety, they can remain operational when isolated from the grid, dispatch power in increments as needed to support the recovery of the grid, and have the ability to resume power generation after a shutdown [55].

Advanced large reactors have also incorporated more passive safety systems in their design. A few examples are the passive cooling system and the high-pressure safety injection with core makeup tanks in the containment of the AP1000 [56]; the core catcher and hydrogen recombiners in the EPR and VVER 1200 [57]; and the passive residual heat removal through the steam generators in the event of a station blackout, and the autocatalytic recombiners installed in the containment of the Chinese HPR1000.

Advanced reactors are also being designed to use more reliable fuel, e.g., extruded metallic fuel, TRISO pellets or pebbles, molten salt fuel, and the rather less mature silicon carbide composite cladding [58]. These are accident-tolerant fuels, specifically designed to prevent radionuclide release and hydrogen formation during accidents [59].

In addition to reducing the emergency planning zone while guaranteeing no radionuclide releases offsite exceeding the limit of effective dose to the public, most of these smaller reactors are housed in underground structures or pools, allowing seismic isolation, aircraft impact protection, and better security [60]. From the security point of view, they can incorporate nuclear material accounting, cyber security, transport security, and sabotage mitigation capabilities in their design [61].

Advanced reactors encompass small and large reactors using gas, liquid metal, or molten salt as a coolant, some produce fast neutrons, and some use sodium as the working fluid in sealed heat pipes, where decay heat removal goes through capillary action and the laws of thermodynamics [62]. These reactor designs can usually remain at atmospheric pressure under high temperatures reducing the likelihood and severity of a loss of coolant accident. Small modular reactors can also have integral configurations, for example, the water-cooled iPWR encloses the major components of the primary coolant system—pressurizer, steam generators, and pumps—within the reactor vessel, thus eliminating loss of coolant and pump shaft break design accidents [60].

The passive transition of decay heat removal from a water-cooled method to air-cooled, in the NuScale SMR shown in Figure 6.3, is a good example of advanced safety features in case of an extended loss of onsite and offsite power.

If power is lost, the vent valves on the reactor head open automatically to vent the steam into the containment. The steam condenses on the inside surface of the reactor vessel and the heat is rejected to the reactor building

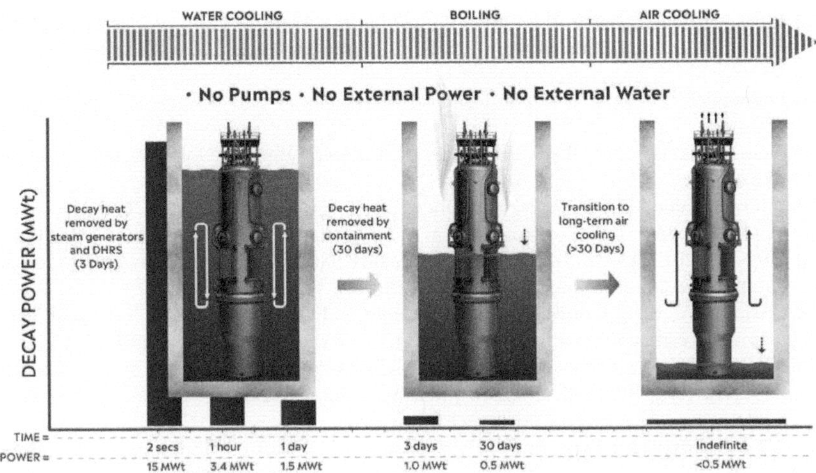

Figure 6.3. NuScale SMR passive transition from water-cooled to air-cooled decay heat removal (Image courtesy of NuScale Power LLC).

pool.[10] When the condensate level in the vessel rises over the core, the recirculation valves open to maintain the core covered with liquid. As seen in Figure 6.3, after a few seconds the core decay power drops. In the next 30 days, if no water is added, the condensate level will decrease from evaporation. At this point, air-natural convection will be used to remove the heat from the external surface of the containment vessel. All this is achieved without operator or computer action, AC or DC power, or the need to add water.

References

[1] Federal Government. 2023. Title 10 Chapter I Part 52—Licenses, Certifications, and Approvals for Nuclear Power Plants. National Archives Code of Federal Regulations, 17 January 2023. [Online]. Available: https://www.ecfr.gov/current/title-10/chapter-I/part-52. [Accessed 20 January 2023].

[2] Federal Government. 2022. Title 10 Chapter I Part 72 Licensing requirements for the independent storage of spent nuclear fuel, high-level radioactive waste, and reactor-related greater than Class C waste. National Archives Code of Federal Regulations, 21 June 2022. [Online]. Available: https://www.ecfr.gov/current/title-10/chapter-I/part-72?toc=1. [Accessed 23 June 2022].

[3] Federal Government. 2022. Title 10 Chapter I Part 71 Packaging and Transportation of Radioactive Material. National Archives Code of Federal Regulations, 1 June 2022. [Online]. Available: https://www.ecfr.gov/current/title-10/chapter-I/part-71?toc=1. [Accessed 7 June 2022].

---

[10] It is the common reactor pool where the reactor vessel is submerged, which serves as the ultimate heat sink for the safety systems.

[4] Euroatom, FAO, IAEA, ILO, IMO, OECD/NEA, PAHO, UNEP, and WHO. 2006. IAEA Safety Standards. Fundamental Safety Principles. Safety fundamentals No. SF-1, Vienna: IAEA.

[5] Domenech, H. 2017. Radiation Safety. Management and Programs, Basel: Springer Nature.

[6] Deitrich, L. W. 2001. Basic Principles of Nuclear Safety (PPT Presentation), Lemont: IAEA.

[7] International Atomic Energy Agency. 2015. Basic Professional Training Course. Module III Basic Principles of Nuclear Safety, Vienna: IAEA.

[8] US NRC. 2012. Module 4.0 Nuclear Criticality Safety 0905 (Rev. 3); Nuclear Criticality Safety Controls, Chatanooga: USNRC Technical Training Center.

[9] Costa, D. A., M. E. Cournoyer, J. F. Merhege, V. E. Garcia and A. N. Sandoval. 2017. A primer on criticality safety. Journal of Chemical Health and Safety 24(3): 7–13.

[10] Nuclear Power for Everybody. 2023. Reactor Criticality. A Group of Engineers, n.d. [Online]. Available: https://www.nuclear-power.com/nuclear-power/reactor-physics/nuclear-fission-chain-reaction/reactor-criticality/. [Accessed 2023 October 2023].

[11] Nuclear Power for Everybody. 2023. Reactivity. A Group of Engineers, n.d. [Online]. Available: https://www.nuclear-power.com/nuclear-power/reactor-physics/nuclear-fission-chain-reaction/reactivity/. [Accessed 24 July 2023].

[12] Bartel, R., T. Wellok and R. J. Budnitz. 2016. NUREG/KM-0010, WASH 1400 The Reactor Safety Study. The Introduction of Risk Assessment to the Regulation of Nuclear Reactors. US Nuclear Regulatory Commission, Washington.

[13] Nourbakhsh, H. P. 2014. Dealing with beyond-design-basis accidents in nuclear safety decisions. In 12th International Probabilistic Safety Assessment & Management (PSAM 12), Honolulu.

[14] International Atomic Energy Agency. 2021. Specific Safety Guide No. No. SSG-64, Protection against Internal Hazards in the Design of Nuclear Power Plants, Vienna: IAEA.

[15] International Atomic Energy Agency. 2020. Safety Reports Series No. 103 Methodologies for Seismic Safety Evaluation of Existing Nuclear Installations, Vienna: IAEA.

[16] International Atomic Energy Agency. 2018. Safety Reports Series No. 92 Consideration of External Hazards in Probabilistic Safety Assessment for Single Unit and Multi-unit Nuclear Power Plants, Vienna: IAEA.

[17] US NRC. 2023. Computer Codes. Nuclear Regulatory Commission, 10 March 2023. [Online]. Available: https://www.nrc.gov/about-nrc/regulatory/research/safetycodes.html. [Accessed 25 July 2023].

[18] Federal Government. 2023. Title 10 Chapter I Part 50 Appendix B to Part 50 —Quality Assurance Criteria for Nuclear Power Plants and Fuel Reprocessing Plants. National Archives Code of Federal Regulations, 24 July 2023. [Online]. Available: https://www.ecfr.gov/current/title-10/chapter-I/part-50/appendix-Appendix%20B%20to%20Part%2050. [Accessed 26 July 2023].

[19] American Society of Mechanical Engineers (ASME). 2022. ASME NQA-1 Quality Assurance Requirements For Nuclear Facility Applications, New York: ASME.

[20] International Atomic Energy Agency. 2022. Nuclear Safety and Security Glossary. Terminology Used in Nuclear Safety, Nuclear Security, Radiation Protection and Emergency Preparedness and Response. IAEA. [Online]. Available: https://kos.iaea.org/iaea-safety-glossary.html.

[21] US NRC. 2021. Nuclear Security and Safeguards. Nuclear Regulatory Commission, 6 October 2021. [Online]. Available: https://www.nrc.gov/security.html. [Accessed 27 July 2023].

[22] Nuclear Energy Institute (NEI). 2016. Nuclear Power Plant Security and Access Control. Fact Sheet. Nuclear Energy Institute, September 2016. [Online]. Available:

https://www.nei.org/resources/fact-sheets/nuclear-plant-security-and-access-control. [Accessed 27 July 2023].

[23] Nuclear Energy Institute (NEI). 2010. NEI 08-09 [Rev. 6]. Cyber Security Plan for Nuclear Power Reactors. Nuclear Energy Institute, Washington.

[24] International Atomic Energy Agency. 2016. Safety Standards Series No. SSR-2/1 (Rev 1), Safety of Nuclear Power Plants: Design, Vienna: IAEA.

[25] Federal Government. 2023. Title 10, Chapter 1, Part 50, Appendix A to Part 50 General Design Criteria for Nuclear Power Plants. National Archives Code of Federal Regulations, 3 July 2023. [Online]. Available: https://www.ecfr.gov/current/title-10/chapter-I/part-50/appendix-Appendix%20A%20to%20Part%2050. [Accessed 2023 10 July].

[26] Federal Government. 2023. Title 10 Chapter I Part 50 General Provisions § 50.2 Definitions. National Archives Code of Federal Regulations, 14 July 2023. [Online]. Available: https://www.ecfr.gov/current/title-10/chapter-I/part-50/subject-group-ECFR1aad221999377bf/section-50.2. [Accessed 18 July 2023].

[27] International Atomic Energy Agency. 2014. Specific Safety Guide No. SSG-30. Safety Classification of Structures, Systems, and Components in Nuclear Power Plants, Vienna: IAEA.

[28] Nuclear Power for Everybody. 2023. Level 2 – Abnormal Operation. Anticipated Operational Occurrences. A Group of Engineers, n.d. [Online]. Available: https://www.nuclear-power.com/nuclear-power/reactor-physics/nuclear-safety/level-2-abnormal-operation/. [Accessed 28 July 2023].

[29] US NRC. 2021. Design-basis accident. Nuclear Regulatory Commission, 09 March 2021. [Online]. Available: https://www.nrc.gov/reading-rm/basic-ref/glossary/design-basis-accident.html. [Accessed 28 July 2023].

[30] International Atomic Energy Agency. 2023. IAEA-TECDOC-2021. Experience in Applying IAEA Principles for Design Safety to New Nuclear Power Plants, Vienna: IAEA.

[31] US NRC. 2021. Defense in depth. Nuclear Regulatory Commission, 9 March 2021. [Online]. Available: https://www.nrc.gov/reading-rm/basic-ref/glossary/defense-in-depth.html. [Accessed 30 July 2023].

[32] International Nuclear Safety Advisory Group. 1996. Safety Series No. 75 INSAG-10 Defense in Depth in Nuclear Safety. IAEA, Vienna.

[33] International Atomic Energy Agency. 2016. IAEA Specific Safety Requirements No. SSR-2/1 (Rev. 1), Safety of Nuclear Power Plants: Design, Vienna: IAEA.

[34] Nuclear Energy Institute (NEI). 2017. White Paper Fundamental Instrumentation and Control Design Principles Rev 0 Draft. Nuclear Energy Institute, Washington.

[35] International Atomic Energy Agency. 2019. Specific Safety Guide No. SSG-53 Design of the Reactor Containment and Associated Systems for Nuclear Power Plants, Vienna: IAEA.

[36] Salem Generating Station. 1985. Updated Final Safety Analysis Report. Section 12 Radiation Protection. Salem Generating Station, Salem.

[37] International Atomic Energy Agency. 2023. PRIS Power Reactor Information System. The Database on Nuclear Power Reactors. IAEA, 23 January 2023. [Online]. Available: https://pris.iaea.org/pris/. [Accessed 24 January 2023].

[38] International Atomic Energy Agency. 1998. International Nuclear and Radiological Event Scale (INES). IAEA. [Online]. Available: https://www.iaea.org/resources/databases/international-nuclear-and-radiological-event-scale. [Accessed 16 June 2023].

[39] UNSCEAR. 2011. Sources and effects of ionizing radiation—UNSCEAR 2008 Report. Volume II Annex D - Health effects due to radiation from the Chornobyl accident. United Nations, New York.

[40] Ritchie, H. 2022. What are the safest and cleanest sources of energy? Our World in Data, 10 February 2022 update. [Online]. Available: https://ourworldindata.org/safest-sources-of-energy. [Accessed 3 August 2023].

[41] Britannica. 2023. The Editors of Encyclopaedia, Bhopal disaster. Encyclopedia Britannica, 30 June 2023. [Online]. Available: https://www.britannica.com/event/Bhopal-disaster. [Accessed 12 August 2023].

[42] World Nuclear Association. 2022. Safety of Nuclear Power Reactors. World Nuclear Association, March 2022. [Online]. Available: https://www.world-nuclear.org/information-library/safety-and-security/safety-of-plants/safety-of-nuclear-power-reactors.aspx. [Accessed 7 August 2023].

[43] US NRC. 2021. Emergency Classification. Nuclear Regulatory Commission, 29 March 21. [Online]. Available: https://www.nrc.gov/about-nrc/emerg-preparedness/about-emerg-preparedness/emerg-classification.html. [Accessed 8 August 2023].

[44] US NRC - US Department of Homeland Security FEMA. 2011. NUREG-0654 FEMA-REP-1, Rev. 1 Supplement 3. Criteria for Preparation and Evaluation of Radiological Emergency Response Plans and Preparedness in Support of Nuclear Power Plants. Guidance for Protective Action Strategies. Nuclear Regulatory Commission, Washington.

[45] International Atomic Energy Agency. 1986. Convention on Early Notification of a Nuclear Accident, Vienna: IAEA.

[46] International Atomic Energy Agency. 1986. Convention on Assistance in the Case of a Nuclear Accident or Radiological Emergency, Vienna: IAEA.

[47] International Atomic Energy Agency. 2023. Incident and Emergency Centre. IAEA, 2023. [Online]. Available: https://www.iaea.org/about/organizational-structure/department-of-nuclear-safety-and-security/incident-and-emergency-centre. [Accessed 4 August 2023].

[48] Incident and Emergency Centre. 2005. IAEA, Answering the Request for Emergency Assistance Worldwide. The Incident and Emergency Centre (IEC), Vienna: IAEA.

[49] FEMA. 2023. Radiological Emergency Preparedness. FEMA, 2 May 2023. [Online]. Available: https://www.fema.gov/emergency-managers/practitioners/hazardous-response-capabilities/radiological#. [Accessed 5 August 2023].

[50] US NRC. 2020. About Emergency Preparedness. Nuclear Regulatory Commission, 4 November 2020. [Online]. Available: https://www.nrc.gov/about-nrc/emerg-preparedness/protect-public.html. [Accessed 5 August 2023].

[51] FEMA. 2022. 2021 Annual Report Radiological Emergency Preparedness Program. FEMA, Washington.

[52] Institute of Nuclear Power Operations (INPO). 2004. Principles for a Strong Nuclear Safety Culture. Institute of Nuclear Power Operations, Atlanta.

[53] Nuclear Power for Everybody. Reactivity Coefficients – Reactivity Feedbacks. A Group of Engineers, n.d. [Online]. Available: https://www.nuclear-power.com/nuclear-power/reactor-physics/nuclear-fission-chain-reaction/reactivity-coefficients-reactivity-feedbacks/. [Accessed 7 August 2023].

[54] Reyes Jr., J. N., F. Southworth and B. G. Woods. 2020. Why the unique safety features of advanced reactors matter. The Bridge 50(3): 45–51.

[55] Hussein, E. M. 2020. Emerging small modular nuclear power reactors: A critical review. Physics Open 5: Art. 100038.

[56] Westinghouse. AP1000 Nuclear Power Plant - Passive Safety Systems. Westinghouse Electric Company LLC, n.d. [Online]. Available: https://www.westinghousenuclear.com/energy-systems/ap1000-pwr/safety/passive-safety-systems. [Accessed 9 August 2023].

[57] World Nuclear Association. 2021. Advanced Nuclear Power Reactors. World Nuclear Association, April 2021. [Online]. Available: https://world-nuclear.org/information-

library/nuclear-fuel-cycle/nuclear-power-reactors/advanced-nuclear-power-reactors. aspx. [Accessed 9 August 2023].

[58] Ultra Safe Nuclear. 2023. TRISO/FCM® FUEL. Ultra Safe Nuclear Corporation, 2023. [Online]. Available: https://www.usnc.com/fuel/. [Accessed 16 April 2023].

[59] US NRC. 2023. Accident Tolerant Fuels. Nuclear Regulatory Commission, 9 February 2023. [Online]. Available: https://www.nrc.gov/reactors/power/atf.html. [Accessed 4 April 2023].

[60] Advanced Systems Technology and Management. 2016. Presentation to the Arab Atomic Energy Agency: SMR: Safety and Security, Vienna: NRC-IAEA.

[61] Department of Energy's National Nuclear Security Administration. 2020. Recommendations and Resources for Advanced Reactor Developers, Washington: Department of Energy (DOE).

[62] Swartz, M. M., W. A. Byers, J. Lojek and R. Blunt. 2021. Westinghouse eVinci™ heat pipe micro reactor technology development. In 28th International Conference on Nuclear Engineering (ICONE), Online.

CHAPTER 7

Environmental Impact of Nuclear Materials

Renewable energy sources such as solar, wind, and hydropower as well as nuclear energy, are clean and efficient ways of producing electricity and heat. Burning fossil fuels emits greenhouse gases—CO_2, methane, and nitrous oxide—that trap solar heat and contribute to climate change. As a result, the world is warming faster than at any point in recorded history [1]. Although fossil fuels account for over 75% of global greenhouse gas emissions and nearly 90% of all CO_2 emissions [2], by now, around 84% of world energy is still produced from fossil fuels, such as coal, oil, and gas, even if the percent varies depending on the type and source of energy.

Nuclear power accounts for about 10% of electricity generation globally, rising to almost 20% in advanced economies like the UK and the US. The countries with more nuclear contribution are France (63%), and Sweden (30%). The global contribution of solar and wind energy is also growing. In 2022, wind and solar generated 12% of the world's power. The countries with more renewable contributions are Brazil (87%), Sweden (68%), and the UK (41%) [3]. Nuclear, hydropower, solar, and wind power do not emit greenhouse gases or pollutants into the air. The graphic in Figure 7.1 provides an image of the contribution of each source to energy consumption in 2022 [3].

According to a report by the International Energy Agency (IAE), the switch from coal to natural gas, which started in 2011, has been a step forward in the reduction of emissions and air pollutants, alongside the rise of renewable wind and solar energy [4]. Following these policies, some countries have reduced greenhouse emissions by a third or more, and some have achieved reduction rates of about 4% a year [5]. However, gas consumption only reduces emissions, it does not eliminate them.

Nuclear energy on the other hand does not produce CO_2 or greenhouse gases. It also operates at a much higher capacity than solar and wind, than even fossil fuels. Deploying more nuclear energy, along with renewables,

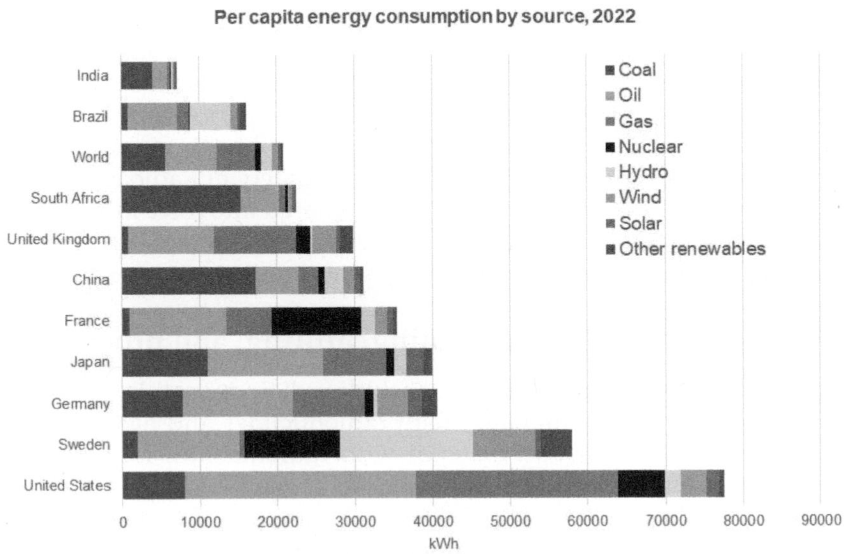

Per capita energy consumption by source, 2022

Figure 7.1. Primary energy consumption per person in the world and selected countries in 2022, by source (Graphic courtesy of Our World in Data).

will significantly contribute to the reduction of greenhouse emissions and will provide short and long-term benefits to the grid [6].

The last has been demonstrated to be certain over the last 60 years. Nuclear power is dependable and can be deployed on a large scale. It is a fact that electricity can be generated by nuclear plants most of the year ($\approx$ 92% of the year), while, e.g., solar systems deliver power only around 27% of the time, natural gas units 57%, and coal plants 48% [7].

Nevertheless, preconceptions remain among the public about the safety of nuclear power plants and their radioactive waste generation, although these still release less radiation into the environment than any other major energy source. For example, coal rock contains small amounts of natural radioactive elements uranium and thorium, and when it burns, produces more radioactive releases into the environment than a nuclear plant producing the same amount of energy [8].

Besides, nuclear power is among the most heavily regulated commercial enterprises in the entire world [9]. Regulations are enforced through licenses, inspections, corrective actions, fines, and even the shutdown of a facility.

Low-level radioactive effluents released to the environment by research and commercial nuclear reactors are regulated by national authorities to ensure they do not exceed dose limits to the public. The IAEA database DIRATA (Discharges of Radionuclides to the Atmosphere and the Aquatic Environment) collects voluntary information from countries to further provide useful information exchange and tracking trends about releases [10]. The database allows selecting the country, site type—e.g., nuclear power plant or other—a release type—i.e., atmospheric or liquid—the nuclide emission, the specific nuclide, and the period for the release.

DIRATA database supports the objectives of the Joint Convention on the Safety of Spent Fuel Management and the Safety of Radioactive Waste Management [11]. This convention, which entered into force in 2001, is the first legal instrument to globally address this issue to promote a high level of safety for the management of spent fuel and radioactive waste [12].

Radioactive waste from mining, milling, and processing

Radioactive waste is any material, including liquids, solids, and gases, that contains or is contaminated with radionuclides and has no further use. However, because any material with radionuclides is radioactive, regardless of its activity or concentration, a material is only classified as radioactive waste if the concentration or activity of its radionuclides exceeds the limit set by the regulatory authority [13].

Waste from mining, milling, and processing includes various types. Solid waste comprises waste rock, mill tailings, and mineralized waste rock.[1] Liquid waste is process water, known as raffinate, and includes solutions from mineral leaching. In addition, used pipes, process vessels, filters, tools, and other discarded solids are considered radioactive waste.

Waste rock—clean and mineralized—is simply rock material removed from the mine to gain access to the ore. It is also called bulk rock because its amount can be very large. Apart from heavy metals, waste rock might contain all the radionuclides in the original ore but at lower activity concentrations, usually below 1 Bq/g.[2] Compared to normal rock, waste rock can still result in public exposure above 1 mSv[3] in a year. Uranium mill tailings are sandy-like tailings or sludges resulting from alkaline or acidic processes to extract the uranium or thorium. Mill tailings are also produced in large quantities and retain around 75% of the original radioactivity present in the ore, along with ^{230}Th, ^{226}Ra, their decay

---

[1] Mineralized waste rock contains either low-grade ore or significant concentrations of secondary minerals.

[2] Since 1975, becquerel is the unit of radioactivity. It is defined as the activity of a quantity of radioactive material in which one nucleus decays per second (1 Bq = 1 disintegration per second). The unit was named in honor of Henri Becquerel, a French physicist who discovered radioactivity in 1896. The old unit was curie (Ci). 1 Ci = 3.7 × 10^{10} Bq = 37 GBq.

[3] This is the effective dose limit for individual members of the public. This means that the contribution of this source to an individual should not add more than 1 mSv to the natural background the individual is already exposed to.

products, and toxic heavy metals such as copper, arsenic, molybdenum, and vanadium [14].

Mill tailings are typically dumped at disposal sites as sludge or slurries, in specially engineered structures called impoundments, above or below ground level, covered with a sealing barrier to prevent radon from escaping. Water cover may be used to reduce radon emanations and prevent problems of acid drainage. Measures for groundwater protection, erosion prevention, inspection, and surveillance are also necessary. The US Appendix A to 10 CFR Part 40 establishes the criteria for the disposition of tailings or waste from natural uranium and thorium [15].

Both waste rock and mill tailing release radon gas and seepage water that might contain radioactive and toxic materials. These also increase the level of gamma radiation in the surrounding area. Hence, mill tailings piled up from nuclear weapon programs, and early nuclear power in the past have led to environmental problems in the sites where they were abandoned [16]. To address these concerns, in 1978, the US Congress enacted the Uranium Mill Tailings Radiation Control Act (UMTRCA) to provide for the disposal, long-term stabilization, and control of uranium mill tailings in legacy sites[4] [17].

The Moab Uranium Mill Tailings Remedial Action Project illustrated in Figure 7.2, is an example. This 2 km^2 site was a former uranium-ore processing facility that operated under private ownership from 1956 to 1984. Here, a pile of around 16 million tons of uranium mill tailings was accumulated in an unlined impoundment close to the Colorado River [18]. The project involved the relocation of all the waste to an engineered disposal site, approved by the NRC, near Crescent Junction, around 48 km north of Moab.

The shipping of the sand-like, mill tails from Moab to Crescent started in 2009, and as of today, more than 13 million tons have already been shipped in locked, steel railcar intermodal containers, to be disposed of in the Crescent Junction disposal cell that was previously excavated to around 8 m deep below the surface.

Some photos of the containers and operations can be seen in Figure 7.3. After transferring the containers from the train to haul trucks, these are driven to the disposal area where the mill tailings are unloaded and scattered. Next, a sheep's foot roller equipped with a computer-based compaction system is used to compact the residual radioactive material (RRM) inside the disposal cell until the final grade for tailings material is met. At both Moab and Crescent Junction sites, water trailers are used to suppress dust and prevent ^{226}Rn dispersion to populated areas.

---

[4] Legacy sites mean sites that produced uranium for the early nuclear power and weapons programs and were closed prior to 1978.

Figure 7.2. The Moab legacy site subject to the removal of waste located near the Colorado River is shown on the left. On the right is the Crescent Junction disposal site where the mill tailings have been relocated (Photos courtesy of the US DOE).

Figure 7.3. (1) The containers are transferred to and from the train using haul trucks. (2) At the Crescent Junction site, the tailings are discharged into the disposal cell. (3) At both sites, water trailers are used to suppress dust and prevent radon from dispersing (Photos courtesy of the US DOE).

The disposal cell is capped with multi-layers of locally sourced soils and rock, as seen in Figure 7.4. The first layer over the mill tails is an interim cover to prevent the escape of gas. A second layer of processed Mancos Shale is used as a barrier to radon release. This layer is covered with sandy gravel to form an infiltration and bio-intrusion barrier. The frost protection layer is made of alluvial and eolian soils and weathered Mancos Shale. Finally, a layer of rock is applied at the top [19].

Generally, a self-sustaining vegetative cover is also employed to reduce wind and water erosion to negligible levels [15]. This disposal cell

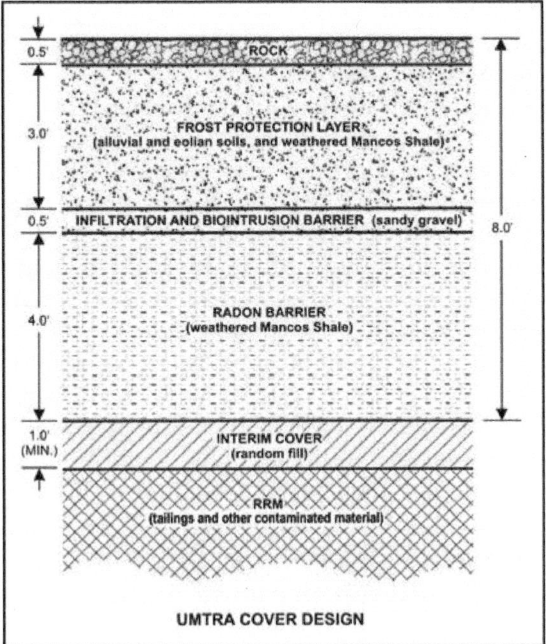

Figure 7.4. Diagram of cover layers for Crescent Junction disposal cell (Image courtesy of the Department of Energy).

should be designed to be effective for 1,000 years to the extent reasonably achievable, with a minimum performance period of 200 years.

Because the soil in the Moab site has absorbed uranium and has been slowly releasing it into the groundwater, in addition to relocating the tailings, there is an active project for remediation of the Moab site's groundwater using extraction and injection wells to protect the Colorado River. This protection consists of removing the contaminant mass— uranium and ammonium—using groundwater extraction wells and injecting freshwater into the underlying alluvium through remediation wells located near the western bank of the river. There are also monitoring wells for surveillance [20].

Despite almost 90–95% of uranium being extracted from the rock by the milling process, the mill tailings contain toxic heavy metals from the rock, and nearly all of the radioactive daughters from the decay of uranium, including radium and radon gas. They can also include liquid waste resulting from uranium solution extraction processes, such as *in situ* recovery, heap leach, and ion exchange. They are all stored in impoundments.

Other radioactive wastes from mining and milling include tools, pipes, vessels, filters, protective clothing, wiping cloths, gloves, and other disposable items that become contaminated with some amounts of radioactive dust or particles and are not allowed to come in contact with individuals and the outside environment.

These low-level solid wastes (LLW) are long-lived (their half-life T½ > 30 years) and, according to their level of activity, in the US, are classified in A, B, or C, and greater than class C (GTCC) waste. GTCC is waste whose concentration of radionuclides exceeds the limits established for Class C and requires deeper disposal.

LLW from mining and milling are first separated from other materials and collected. Then, they are classified by activity as indicated in the above paragraph, segregated as compactable or combustible, treated accordingly to reduce their volume, and stored on site until amounts are large enough for shipment to be buried at licensed facilities. In the US, such radioactive waste shallow burial facilities operate in Washington, Utah, Texas, Tennessee, and South Carolina. These facilities are designed, constructed, and operated to meet safety standards [21].

Radioactive waste from fuel fabrication and reprocessing plants

Waste generated in the conversion and enrichment of uranium, and the fabrication of fuel elements, includes a variety of nuclear waste materials such as depleted uranium, uranium and plutonium oxides, other fission products, and so forth; as well as toxic chemicals such as fluoride acid (HF), nitrates, etcetera. These wastes are in gaseous, liquid, and solid forms.

Conversion is the stage of the nuclear fuel cycle where the triuranium octoxide (U_3O_8) is converted into UF_6, the form of uranium required for enrichment. In this process, the U_3O_8 is first purified, reduced to UO_2, and then combined with fluorine using strong chemicals such as fluorine, hydrofluoric acid, and uranyl fluoride. This process only produces waste with a low (LLW) or very low level of radioactivity. It is noteworthy that conversion plants are operating commercially only in Canada, France, Russia, and China [22].

Produced waste can be in gaseous, liquid, and solid forms. Gaseous streams from the production processes, and air renovation of building ventilation systems, are filtered and uranium-contaminated filter materials are segregated and treated as solid waste, together with slightly uranium-contaminated trash and residues from fluorine recovery operations. The resulting gaseous effluents are discharged into the atmosphere after being monitored to ensure the corresponding limits are met.

Liquid streams may be processing effluents, cooling water, rainwater, sanitary wastewater, or the like. These are sent to the liquid waste treatment workshop to be decontaminated and chemically neutralized.

There, radioactive impurities are precipitated and trapped as solid waste, and the resulting liquid effluents are filtered and monitored before their release. Each release is only completed after rigorous sampling and monitoring by the radiation protection unit to ensure established limits are met. Released volumes and amounts are recorded and reported as required by the regulatory authority.

Solid waste, mainly consisting of the above-mentioned filter materials, contaminated trash (used gloves, protective clothing, filters, samples, paper tissues, discarded tools, and so on), precipitates from liquid treatment, residues from recovery and cleaning operations, etc., is finally sorted, classified, and packaged to be temporarily stored before being dispatched to the appropriate waste facilities for the disposal of low-level radioactive waste.

Therefore, the management of radioactive waste encompasses first its specific segregation and classification—by type, half-live, and level of activity—then, its pretreatment, treatment, and conditioning to reduce its volume, and finally its storage and transport to a disposal facility.

Because uranium and its decay products are long-lived radionuclides ($T\frac{1}{2}$ > 30 y) they will remain hazardous for thousands of years and need to stay stored for extended periods. In the US only low-level waste of classes A, B, and C are allowed at shallow burial commercial facilities. GTCC (greater than C) uranium waste would require more stringent methods of disposal to provide additional measures of isolation from the environment and protection against intrusion (geological repositories are addressed forward) [23].

The enrichment process generates large amounts of radioactive waste in the form of tails of depleted uranium hexafluoride (DUF_6). Typically, these tails are transferred to storage cylinders and stored in open-air backyards. For example, around 800,000 tons of such tails are stored at the Paducah and Portsmouth sites in the US.

As was noted before, DUF_6 is an unstable compound that retains its radioactivity for very long periods. Although external exposure is not considered a serious hazard since depleted uranium is an alpha emitter, in the case DUF_6, if released as a result of container corrosion or other damages, ingested or inhaled, depleted uranium can cause severe damage to individuals [24].

The deconversion of these tails can significantly reduce their chemical hazards by extracting the fluoride atoms and replacing them with oxygen to produce DUO_2 and DU_3O_8. Both these uranium oxides are chemically stable, can be safely stored, and are better suitable for disposal as low-level radioactive waste. Deconversion also enables the recovery of high-purity fluoride compounds that have commercial value. For example, Orano deconverts DU into U_3O_8 via the defluorination process in France [25], and Urenco is responsible for deconverting UF_6 to U_3O_8 in the UK [26].

Depleted uranium could also be recycled. That is, re-enriched. Orano is considering the economic viability of obtaining enriched natural uranium fuel from depleted uranium hexafluoride mostly to natural-equivalent grade, but also to reactor grade for use in existing and IV-generation reactors [27].

Waste generation is one of the arguments used against nuclear power, however, one of its benefits is that the used fuel can be recycled instead of being discarded as waste, thus closing the fuel cycle. This process is known as nuclear fuel reprocessing and mainly involves the recovery of uranium and plutonium from the spent fuel. According to Orano, in the La Hague plant, a leading reprocessing plant in the world, 96% of nuclear material is recovered, and only 4% goes to the final conditioned waste [28].

Currently, there are only three countries in the world with reprocessing facilities: France, India, and Russia [29]. Japan is still performing commissioning tests in its planned Rokkasho plant for spent fuel reprocessing. The Thermal Oxide Reprocessing Plant (Thorp) in the UK ceased reprocessing in 2018. In the US, the existing reprocessing plants have been for military use exclusively.

Reprocessing operations generate solid, liquid, and gaseous radioactive waste streams. The gaseous waste stream is generated as off-gas in the different processes of the spent fuel and vitrification of waste. These might contain noble gases such as ^{85}Kr (krypton), ^{41}Ar (argon), ^{133}Xe (xenon), ^{129}I, ^{14}C, and ^{3}H. These can also contain alpha and beta fission products (^{90}Sr, and ^{137}Cs) from the fuel. Off-gases are first siphoned and treated with a series of scrubbers and absorbers to remove the nitrogen oxides. Finally, gases pass through high-efficiency filters to remove alpha and beta aerosols before discharge into the atmosphere. Scrubbers, absorbents, and filters will be discharged as low-level or intermediate-level radioactive solid waste [30].

Liquid waste from reprocessing is high-level, intermediate-level, and low-level radioactive waste. These wastes are produced at the different stages of fuel reprocessing. High-level liquid radioactive waste (HLW)[5] results from the removal of plutonium and unburned uranium during the first step of spent fuel dissolution. It generates decay heat and contains minor actinide[6] elements that were produced during fuel burnup, in

---

[5] It is important to notice that, by volume, HLW represents only around 1% of all waste, while, in activity, it is around 95%.

[6] Minor actinides include neptunium, americium, curium, berkelium, californium, einsteinium, and fermium. Uranium and plutonium are major actinides.

high concentrations, as well as small amounts of residual uranium and plutonium. Liquid HLW is conditioned in glass by a vitrification process to facilitate its long-term storage (tens of thousands of years). Through vitrification, the liquid waste is first concentrated in an evaporator to reduce its volume, then dried in a calciner, mixed with borosilicate glass, and melted in a metallic container where it solidifies after cooling [31].

Intermediate-level radioactive liquid waste (ILW), with lower activity and lower heat generation, results from the rest of the operations in the reprocessing process. These include scrubber solutions, ion-exchange slurries, chemical sludges, decontamination and wash liquids, etcetera. To reduce its volume and confine it to a form more suitable for handling, storage, and disposal, ILW is conditioned in glass or concrete [32]. ILW is similar to types B and C in the US.

Very-low-level liquid wastes (VLLW) and low-level liquid wastes (LLW) are collected from drainage systems for rainwater in specific zones — e.g., on the storage casks — water containing chemicals from laboratories, water from installations including a spent fuel storage pool, solutions from repair and maintenance, and domestic water from buildings. LLW liquids, similar to type A in the US, are typically concentrated to reduce their volume by ion exchange or chemical coprecipitation. The resulting condensates and sludges are finally conditioned with cement. Post-treatment, cleared fluids can be usually released as effluents after the corresponding measurement and control [33]. VLLW liquid waste can also be released with liquid effluents in many cases.

In addition to the containers of vitrified HLW, drums with bituminized ILW, and drums with cemented LLW that were obtained in the treatment of liquid waste, solid waste includes fuel element cladding hulls and end-fittings, hardware, and other insoluble pieces from the fuel dissolution with nitric acid. To reduce its volume, these solid ILWs are super-compacted into a disk shape. The compressed disks are then inserted in universal canisters with identical geometries to that of the vitrified liquid waste and grouted with cement. There is also technological waste, that is, waste resulting from the operation and maintenance of plants, such as structural materials and equipment, resins, filters, and personal protective equipment. This waste is typically LLW and is cemented in fiber-concrete containers [34]. Combustible LLW from operations and laboratories can be incinerated to reduce its volume.

Conditioned long-lived waste is currently meant to be stored for many years before it undergoes further processing and final disposal. Therefore, all waste packages[7] are to be standardized and approved by the regulatory

---

[7] The combination of the solidified waste form and its container, which must be suitable for handling, storage, transport, and disposal according to waste acceptance criteria.

authority. Generally, LLW are intended to be disposed of in near-surface disposal facilities, while HLW and ILW are intended to be disposed of in deep geological repositories. The purpose in both cases is to isolate the radioactivity content from the environment using engineering barriers and protect it from inadvertent intrusion [13].

Deep geological repositories are in progress in Sweden, Finland, France, and the US. In France, for example, current waste is stored onsite at La Hague in dedicated rooms pending its transfer to the deep geological storage facility being developed through the CIGEO project [35]. Also, in Sweden, the construction of the SKB final repository for spent nuclear fuel at Forsmark was approved by the government in 2022 and it is expected to be finished in ten years [36]. In the US, the Waste Isolation Pilot Plant (WIPP) is the only licensed geological repository built in salt beds for the disposal of long-lived, non-heat-generating waste from defense activities [37].

It is worth noting that the Swedish system for radioactive waste also comprises two more facilities: the final repository for short-lived intermediate and low-level waste at Forsmark, located in bedrock about 50 meters below sea level, which started operating in 1988 and was the first of its kind in the world [38]; and CLAB, the Central Interim Storage Facility for Spent Nuclear Fuel operating from 1985, which stores the spent fuel in pools located in rock vaults at 25–30 meters underground [39].

In Finland, the Onkalo repository, whose underground cave is shown in Figure 7.5, is close to being the first permanent disposal site for high-level nuclear waste in the world. The idea behind Onkalo is the Swedish multi-barrier principle, where the first barrier is the fuel assembly (fuel pellet, fuel rod, and assembly); the second and third barriers are the inner spheroidal graphite cast-iron and the outer copper canisters—argon is to be injected between the two canisters and sealed tightly; the fourth is the vertical disposal tunnel for the cask, which is also filled with bentonite buffer and backfilled with material when the hole is full; and the final and fifth barrier is the granite bedrock that is more than 400 meters deep. There is also an encapsulation plant on the surface connected to the underground disposal by a lift, with which the canisters will be transported [40]. The repository is expected to start operations in 2024.

Two of the first four pilot deposition holes bored in 2022 can be seen in Figure 7.5. As part of the commissioning of the disposal facility, Posiva Oy, the constructor and operator of the Onkalo project, is carrying out a trial run of the final disposal. This is a full-scale disposal test using the final equipment under actual conditions, with the only difference being that the canisters do not contain spent fuel. Those test canisters are equipped with heating elements simulating the decay heat generated by the spent fuel. About 500 sensors are already deployed to sense temperature changes and pressure of the canisters, deposition holes, and surrounding bedrock [41].

Figure 7.5. Pilot holes in a tunnel at the last level of Onkalo spent fuel repository (Photo Kallerna Creative Commons BY-SA 3.0).

Radioactive waste from nuclear reactors

During the operation of nuclear reactors, gaseous waste can result, e.g., from leakages in the coolant systems, the moderator systems, or the reactor itself. Some activation or contamination could also occur in the ventilation air, e.g., noble gases, nitrogen, oxygen, argon, and tritium. Volatile gases may be released from the spent fuel storage or spent fuel handling operations [42]. Radioactive particulate contaminants and aerosols in the air are usually removed by filtration using HEPA[8] filters. Charcoal filters or sorption beds charged with activated carbon are also used for iodine and noble gases. Prefilters and double filters might be used too. In addition, there might be scrubbers for the removal of gaseous chemicals, particulates, and aerosols from off-gases [43].

The primary sources of liquid waste in the operation of water-cooled nuclear reactors are the primary coolant and the water from the spent fuel storage pools. These could be contaminated with radionuclides from the fuel cladding surfaces, the activation of materials surrounding the fuel, and tritium formed because of the neutron absorption by boron, lithium, or deuterium [44].

---

[8] High-efficiency particulate absorbing filters.

Liquid waste can also result from failures or leakages, runoffs from equipment, floor, and chemical drains; wastewater from laundry, showers, and sinks, and drains from decontamination and maintenance areas, among others. Condensate cleaning and spent fuel pool cleaning are also sources of liquid waste. In addition, water treatment may generate wet solid waste in the form of slurries and sludges.

Solid radioactive waste comprises a variety of items, from spent ion-exchange resins, evaporator, demineralization, and flocculation concentrates, cartridge filters and precoat filter beds, charcoal beds, filters, and scrubbed beds from ventilation and off-gas systems to contaminated scraps, rags, clothing, paper, plastic, and the like, and contaminated equipment, tools, deteriorated parts, core components (control rods, drums, etc.), and debris from fuel assemblies or other in-reactor components.

Although wastes cannot be prevented because they are an intrinsic part of good practices and safe working conditions, they should be minimized by applying specific management measures. Figure 7.6 schematically represents the general approaches and stages of a system for the management of radioactive waste.

The minimization of waste starts at its origin. To begin with, radioactive waste is collected and kept apart from exempt or nonradioactive residual materials which are handled as common waste. There should be also pretreatment proceedings aimed at reducing the amount of waste. The first step is characterization, which allows wastes to be segregated into specific streams by physical state, half-life, and activity, though they are handled similarly.

One criterion is the clearance level[9] for reuse or discharge. Following this criterion, some wastes with $T\frac{1}{2} \leq 120$ days can be stored for a certain period to wait for decay to reach the clearance level to be discharged [45]. In such cases, the waste should be packaged, labeled, and stored consistent with written procedures in agreement with existing regulations [46]. Delay and decay is a valid approach, used for many years now. It involves storing the waste for at least 10 $T\frac{1}{2}$ during which the initial activity will be reduced approximately 1000 times.

Likewise, these wastes can be further segregated according to the method of treatment, conditioning, or storage destined for. For example,

---

[9] The clearance levels are values of activity concentration and/or total activity established by the Regulatory Authority, at or below which a radioactive waste or material may be released from regulatory control.

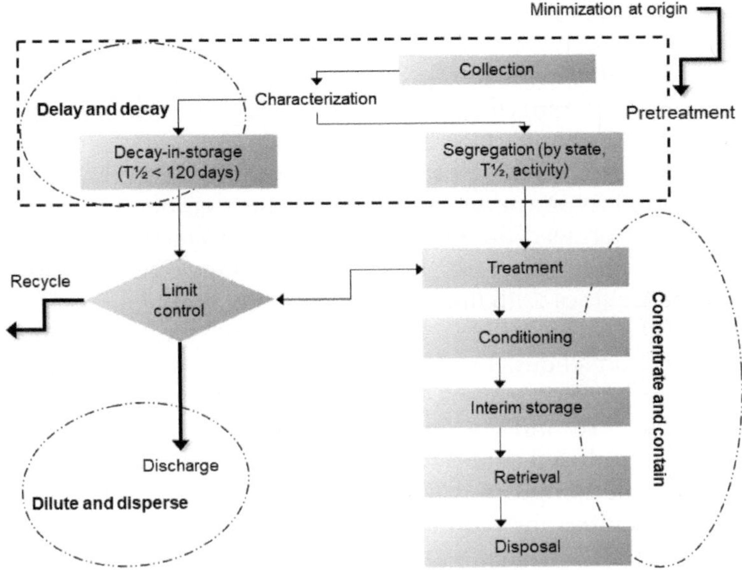

Figure 7.6. Basics stages and approaches for the management of radioactive waste.

solid waste can be segregated as compressible or combustible and sent to the corresponding process of compacting or incineration for volume reduction. Some non-combustible, non-compressible waste, if feasible, can also be shredded. If can be decontaminated, certain solids, such as machines and tools, may be cleaned to the appropriate limits and sent back to the process. Liquid waste can also be segregated depending on specific further treatment methods, e.g., aqueous, corrosive, bearing chemicals or detergents, etc., as well as additional hazards, e.g., toxic, flammable, etc.

Most gaseous waste reaches acceptable levels by passing through the high-efficiency filtering and adsorbing systems of the reactor. Once these levels are confirmed to comply with clearance limits, the waste can be discharged into the atmosphere for further dilution. Similarly, aqueous liquid waste that meets clearance levels after treatment, can be released into the effluent flow for dilution and dispersion, or, if the level of purity is high enough, it can be recycled to the operations as makeup water. A large portion of this aqueous liquid waste is recovered.

Segregation implies diverse waste streams for solids, liquids, and gases. Also, for different activity levels and half-lives. The first criterion is the radionuclide half-life. Hence, waste is usually classified as long-lived waste (T½ > 30 years) and short-lived waste (T½ < 30 years) depending on the method that will be used for its disposal. However, they can be segregated into other half-life groups in between them if necessary. The level of radioactivity is also one of the most important criteria. We already learned that they can be VLLW, LLW, ILW, and HLW depending on the activity and the radionuclides involved [47]. In the US, they are A, B, C, and GTCC, considering the concentration of long-lived and short-lived radionuclides for near-surface disposal [48].

Very low-level waste (VLLW) is waste with limited hazard and an activity level of just one or two orders above the exemption level. This type of waste does not need the most stringent requirements of containment and isolation and, therefore, is suitable for disposal in near-surface landfill-type facilities with limited regulatory control.

Low-level waste (LLW) is waste with limited amounts of long-lived radionuclides. The allowable average activity concentration of long-lived beta and/or gamma-emitting radionuclides, such as ^{14}C, ^{36}Cl, ^{63}Ni, ^{93}Zr, ^{94}Nb, ^{99}Tc, and ^{129}I, can be up to tens of kBq per gram. Almost all the waste produced in reactors is LLW. It has the larger volume and the most variable composition. LLW is suitable for disposal in engineered near-surface facilities and requires containment and isolation for hundreds of years.

Intermediate-level waste (ILW) is waste that contains long-lived radionuclides in quantities that require a superior degree of containment and isolation than that provided by near-surface disposal. A depth from a few tens to a few hundred meters is adequate for ILW if both natural barriers and engineered barriers are selected properly. The likelihood of human intrusion is considered to be significantly reduced at such depths.

High-level waste (HLW) is waste that generates significant amounts of heat, contains large concentrations of both short and long-lived radionuclides, and requires the highest degree of containment and isolation to ensure long-term safety. Such containment and isolation could generally be provided by disposal in deep stable geological formations, with engineered barriers. Heat dissipation has to be taken into account in the design of the disposal facility.

Class A waste is waste that is usually segregated from other waste classes at the disposal site. Class B is waste that must meet more rigorous requirements on the waste form to ensure stability after disposal. Class C waste is waste that not only must meet more rigorous requirements on the waste form to ensure stability but also requires additional measures at the disposal facility to protect against inadvertent intrusion. There are also limits of activity for near-surface disposal concerning long-life radionuclides such as ^{14}C, ^{99}Tc, ^{129}I, and ^{14}C, ^{59}Ni, and ^{94}Nb in metals activated within the reactor. Also, more restrictive limits for alpha-emitting transuranic nuclides with half-life > 5 years, ^{241}Pu, and ^{252}Cm.

Class A also has restrictions on the activity of radionuclides with T½ < 5 y, ^{3}H, and ^{60}Co. In addition, classes B and C have restrictions on the activity of waste containing short-life radionuclides like ^{63}Ni, ^{90}Sr, and ^{137}Cs for near-surface disposal. Waste not suitable for near-surface disposal, meaning GTCC, requires a different form and disposal methods, which are generally stricter [49].

Treatment methods for dry solid radioactive waste include press compaction, incineration, and shredding. Before interim or final disposal, compacted waste and shredded waste are grouted with cement to fill empty spaces and reinforce the outer container. Similarly, ashes from incineration and wet radioactive waste—sludges, slurries—are usually conditioned and stabilized by embedding them in bitumen or concrete. Compressed solid high-level waste (HLW) is typically overpacked in concrete containers for shielding and cooling. Nevertheless, these packages should comply with the waste acceptance criteria from the predisposal storage or disposal facility.

The majority of liquid waste from nuclear reactors is LLW and the methods commonly used for treatment are coprecipitation; flocculation; filtration; demineralization; reverse osmosis; evaporation; and ion exchange/adsorption, typically in combination. For example, filtration may be preceded by precipitation or flocculation, while reverse osmosis or ion exchange may be preceded by precipitation or flocculation and filtration. Ion exchange sometimes is combined with carbon adsorption. The idea behind these methods is to concentrate the radioactivity in a smaller volume and to purify the liquors as much as possible so they can be recycled. Liquid organic waste can be incinerated to reduce its volume. The resulting wastes are ashes (solid).

Concentrated sludges, filter coats, absorbers, or membranes are further treated as wet solid waste or recovered for further treatment. Typically, they are embedded in a matrix of cement or bitumen, i.e., conditioned and stabilized, before packaging in a steel drum, or other engineered container, for predisposal or final disposal. Like in the case of solid waste, the final packages of solidified waste should comply with the waste acceptance criteria from the facility. As a safe measure, waste packages are labeled with the radioactive sign and indicate the most important radionuclides within according to their emission—alpha, beta, gamma—half-life, and relative quantity.

The retrieval step in radioactive waste management involves recovering waste packages from either short-term or long-term storage. It may include reconditioning and packaging. This retrieval may be for inspection, final disposal, or further storage in new facilities.

At present, common disposal options for VLLW, LLW, and ILW are near-surface storages, with or without engineered barriers, at depths of tens of meters below ground, or ground level. Below are some examples of repositories in France, Spain, Sweden, the UK, and the US.

In France, the CSA (Centre de Stockage de l'Aube) and the CIRES (Le Centre Industriel de Regroupement, d'Entreposage et de Stockage) are two of the three near-surface facilities for the disposal of LLW and ILW owned by the French National Agency for Radioactive Waste Management (Andra). The third disposal facility is the CSM (Centre de Stockage de la Manche), currently in the closing phase. CSM was the first near-surface repository in France and operated for 25 years. It was also a reference for the design of the subsequent facilities. The CSA took over the disposal activities of the CSM. It has a license for the disposal of $1 \cdot 10^6 \, m^3$ of LLW and ILW, short-lived waste packages. CIRES is a VLLW near-surface repository that also manages the waste out of nuclear power plants (NPP) and large nuclear research facilities [50].

El Cabril, in Spain, is a near-surface disposal facility designed as a permanent solution for the storage of LLW, ILW, and VLLW. Most of VLLW comes from the decommissioning of nuclear power plants. El Cabril has a separate building area for the treatment, conditioning, and control of waste from medical, research, and industrial institutions that are not NPP, and two disposal areas. The one for low-level and intermediate-level disposal comprises two platforms with 16 and 12 structures (disposal vaults). Its projected containment is for 300 years. The other is for very low-level waste and is projected for 60 years. The disposal platforms are equipped with engineered and natural barriers to confine the waste [51].

In Sweden, the Final Repository for Short-Lived Radioactive Waste (SFR) is located at Forsmark, 50 m deep in the rock below the Baltic Sea. It consists of four long rock vaults of 160 m each for LLW and a 50 m high concrete silo, surrounded by bentonite, for ILW. It was designed for $63,000 \, m^3$ of waste. The containment is projected for at least 500 years. Most of the waste already lying there comes from NPP, but also from hospitals, veterinary and medicine practices, research, and industry [38, 52].

The UK Low-Level Waste Repository, near the West Cumbria coastline, is part of Nuclear Waste Services, an organization focused on the management of radioactive wastes. This is a national near-surface repository with engineered concrete vaults for the disposal of solid low-level wastes both fissile and non-fissile. It receives waste already compacted, in large metal ISO freight containers, from the nuclear industry, defense, non-nuclear industries, and educational, medical, and research establishments. The LLW in drums and boxes is first super compacted to minimize volume. The container with compacted wastes is then grouted before disposal in engineered concrete vaults [53].

In the US, there are four active, licensed low-level waste disposal facilities, located in Barnwell, South Carolina; Richland, Washington;

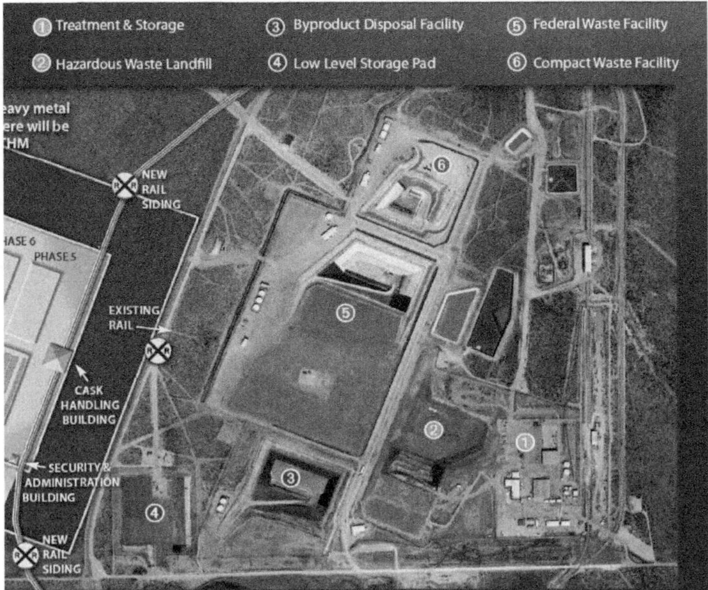

Figure 7.7. Planning for a Consolidated Interim Storage in Andrew, Texas (Photo courtesy of the Nuclear Regulatory Commission).

Clive, Utah; and Andrews, Texas [54]. These disposal sites are designed to confine and isolate Class A, B, and C radioactive wastes [55].

Figure 7.7 shows the facilities in operation in the consolidated interim storage in Andrew, Texas. They are compact waste facilities for the disposal of Class A, B, and C low-level waste from Texas, Vermont, and other states that do not have access to a compact disposal facility (6); the federal waste disposal facility for waste from the Government (5); the byproduct disposal facility where the waste from DOE's Fernald site was disposed of (this is a below-grade disposal site on a thick bed of clay, with an engineered liner, associated support structures on the surface, and a buffer zone) (3); and the hazardous/TSCA/exempt waste facility for the treatment, storage, and landfill disposal of hazardous/toxic waste and exempted radioactive waste (2). The storage pad (4) is an additional storage area [56].

The LLW shallow land burial at Barnwell, South Carolina, is designed for the disposal of concrete rectangular or cylindrical vaults holding sealed waste containers, in trenches excavated in clay-rich soil. When a vault is full, it is closed with a concrete lid. Sandy clay is added as backfill material to fill voids between vaults. Finally, the completed trench area is covered with an engineered cap consisting of a compacted clay layer, a geosynthetic clay liner, a high-density polyethylene (HDPE) liner, a sand layer, and a sandy topsoil layer [57]. Clive, UT, is an above-grade disposal facility limited to Class A, Mixed Waste,[10] NORM, discrete sealed sources, and tailings from the extraction or concentration of uranium or thorium. This site operates both bulk and containerized disposal [58, 59].

It is important to notice that the period of institutional control is always dependent on the radionuclides present in the waste and the site conditions. It can range from several decades to over a hundred years for VLLW or a few hundred years, often 300 years for LLW. Besides, several barriers isolate and contain the waste. Typically, a 2 mm thick high-density polyethylene geomembrane;[11] the natural clay surrounding the excavation; a compacted clay layer with a minimum thickness of 1 m over the filled trenches; a clay backfill of 2.5 m thickness as a minimum to protect the sealed trenches; and a final 30 cm thick vegetative cover to act as a stabilizing layer [60].

Spent nuclear fuel and high-level radioactive waste management

Spent fuel is waste containing higher levels of radioactivity, produces appreciable decay heat, and is long-lived—hundreds of thousands of years. Both current methods of storing and disposing of radioactive waste isolate it from the accessible biosphere by passive engineered and natural barriers but differ in the intention of retrieval.

While disposal means no intention of retrieval, it still does not mean that is not possible. For example, Sellafield, in the UK, was recently the first to make retrievals from the Pile Fuel Cladding Silo as part of site decommissioning. This silo was designed as a locked vault in the 1950s and, after decades in the making and the construction of six large shielded doors, a state-of-the-art robotic arm was recently used to remove and repackage the ILW stored there [61].

The most important approaches for spent fuel interim storage—that is, wet and dry—were discussed in a previous chapter. Both are intended to reduce heat decay and prevent it from adversely affecting the future disposal system. In addition, the experiences of Sweden and Finland in the design and construction of deep geological repositories for disposal were also reviewed.

---

[10] Low-level radioactive waste that also contains components that are chemically hazardous.

[11] A geomembrane is a low-permeability synthetic membrane used as an integral part of a system designed to limit the movement of liquid or gas in the system.

Argillaceous sediments (clays), hard crystalline basement rocks (granite), evaporite formations (dome rock salt), and welded volcanic tuff are the four basic rock types that have been researched for the geological disposal of radioactive waste. However, factors specific to each site should also be considered, e.g., the flow and chemistry of deep groundwater and the mechanical and strength characteristics of the rock [60].

Accordingly, other developments in radioactive waste long-term storage capacities are worth mentioning. They are the Wolsong Radioactive Waste Disposal Centre in South Korea (WLDC); the Center for Geological Disposal, CIGEO, in France; and the Geological Repository Yucca Mountain in the US.

The Wolsong Low and Intermediate-Level Radioactive Waste Disposal Center (WLDC), operating since 2015, is the Korean long-term storage of LILW. This unique repository comprises six silos for ILW storage, located about 80–130 m below the surface, in intrusive igneous rock [60]. The silos are connected to the surface through tunnels. The waste is received in drums which are double-packed in concrete containers at the site. Each one of the silos can hold 1000 of them. The silos are made of 1.2 m thick concrete. The void spaces are planned to be finally backfilled with crushed rock, while concrete plugs will be used to close the accesses. In addition, a near-surface engineered vault-type system for the disposal of LLW is currently under construction, after a review to accommodate a seismic event of magnitude 7.0 [62]. It covers an area of approximately 120,000 m^2 and will handle 1,250,000 two-hundred liter drums of waste [63].

France is one of the three countries that have made the most progress regarding the disposal of radioactive waste. At the beginning of 2023, after 20 years of studies at an underground research laboratory at a depth of 490 m, Andra applied to the regulatory authority for the construction license for CIGEO, its project for a deep geological disposal facility, which is expected to be commissioned in 2035–2040. This facility, the third of its kind worldwide, will be built in an underground area of 30 km^2 selected in the departments of Meuse and Haute-Marne [64].

The underground facility in the stable Callovo-Oxfordian clay formation is projected at a depth of around 500 m, comprising surface facilities for the receipt and conditioning of the waste; a funicular for its transfer underground with an approximately 4.2 km long ramp; access shafts; a disposal area for ILW-LL, and a disposal area for high-level vitrified waste. Technical measures have also been included for the possible retrieval of waste packages in 100 years. Some of these measures include linings to limit cell deformation, robots for retrieving packages, sensors to monitor changes, testing, and so on [65].

According to geological studies, the great majority of radionuclides in the waste will travel only very little within the geological formation and will not rise to the surface over a million years. Andra submitted the CIGEO construction license application to the French Nuclear Safety Authority in January 2023 [66].

The Yucca Mountain repository is the one proposed in the US for the disposal of spent nuclear fuel and HLW. It will use a tunnel approximately 300 m below ground level [67]. Yucca Mountain site consists of a hard, welded volcanic tuff, with a very deep groundwater table—about 500–700 m depth. This project applied for a construction license in 2008 with a design featuring several engineered barriers. A license application for construction was completed for the project but it did not continue for political reasons and was defunded in 2010.

The Yucca Mountain design strongly relies on engineered barriers. The waste packages are placed in approximately 99 parallel linear drifts of an average length of 605 m using an in-drift concept where the waste is placed in steel 2 m × 6 m canisters, nestled in those reinforced storage tunnels [68]. These tunnels are equipped, for example, with a pallet of stainless steel for in-drift support, an inside lining made of sheets of 3 mm stainless steel, and a drift shield made of 15 mm of titanium, along with outer and inner structural shells for the canisters. This design emphasizes the performance of the waste package and the engineered barrier system acting together with the natural unsaturated host unit for the required containment [60].

Decommissioning of nuclear facilities

Decommissioning is the process of retiring from service a nuclear reactor or other nuclear installation when it reaches the end of its life or for other reasons.[12] When the operating license of a reactor should end for any of those reasons, the licensee must close the facility and reduce to safe levels the residual radioactivity in the site so that the regulatory authority can release the property. Therefore, decommissioning is a full process that includes the shutdown of the facility, the removal of nuclear material, the dismantling of systems, structures, and components, the decontamination to reduce or eliminate the residual radioactivity, waste management, and site restoration [69].

The removal of nuclear material implies taking the nuclear fuel out of the reactor and the cooling pool. After that, it undergoes wet or dry interim storage, reprocessing, or storage pending disposal in deep geological repositories according to the country and its nuclear cycle policy. Each

---

[12] The reasons for closing nuclear power plants could be various in different countries: aging, economic considerations, public safety concerns and protests, government political decisions, etcetera.

country also has a specific schedule for these operations. As for the facility systems, structures, and components, it is crucial to distinguish between those located within the neutron radiation field, like the reactor vessel, and the reactor internals; and those contaminated due to their role in the process, like the steam generators, and circulation pumps [70].

The waste arising from the dismantling of the systems, structures, and components within the radiation field includes solids containing long-lived radionuclides from activation. The minor part of this waste is classified as ILW, and the larger part is LLW. Large volumes of VLLW can be also produced. Below is an example of the typical radioactive waste volumes from light water reactors after 40 years of operation [71]:

Reactor type	Radioactive waste volume by class, m^3		
	Class A	Class B/C	GTCC
PWR	6797	184	11
BPR	13,903	372	7

ILW from concrete segmentation typically is the result of dismantling the radiation biological shield of the reactor. Some ILW from metal and concrete segmentation arises from the dismantling of the structures immediately adjacent to the reactor. Radionuclides found in demolition debris and metal scraps with $T\frac{1}{2} > 10$ y are ^{14}C, ^{60}Co, ^{59}Ni, ^{63}Ni, ^{94}Nb, ^{108m}Ag, ^{3}H, ^{36}Cl, ^{41}Ca, ^{151}Sm, ^{152}Eu, and ^{154}Eu. Radionuclides frequently found in contaminated equipment and protection structures are ^{137}Cs, ^{60}Co, ^{134}Cs, $^{90}Sr+^{90}Y$, ^{110m}Ag, and ^{54}Mn. They are mostly gamma and beta emitters [72].

Low volumes of ILW and long-lived LLW resulting from decommissioning are managed via treatment, conditioning, packaging, and disposal as discussed earlier. Treatment involves concentration by the appropriate method to reduce the volume; conditioning implies its immobilization in a borosilicate matrix and, finally, the packaging is designed for interim storage and eventual disposal in a deep repository.

The larger volumes of waste from decommissioning are LLW and VLLW. They include demolition rubble, metal scraps from dismantling, large or cut pieces of metal structures, and soils. LLW and VLLW can be cleared and released into a conventional landfill for disposal if the activity levels show they meet the established clearance levels. If the waste does not satisfy the limits for clearance, but satisfies the limits for recycling or reuse, it is sent back to the nuclear, metal, or construction industry to be further mixed and diluted. For example, concrete or sand can be used as filling material for road construction, metal ingots can be melted together with other metals to produce source material for the metal industry, and soil can be used as filling for site remediation. Otherwise, the waste can be disposed of in facilities for non-nuclear hazardous waste [73].

Generally, radioactive waste is treated or stored within the facility being decommissioned. However, it can also be done in an external facility. Usually, there will be a combination of both. To further minimize the waste, huge equipment, such as steam generators, the core shroud,[13] vessels, tanks, etc., are subject to decontamination too. Concrete can also be decontaminated based on the penetration of the contamination.

Although the majority of wastes from decommissioning are solid, liquid, and gaseous waste can also be produced. According to the segmentation technology chosen, aerosols, dust, and water pollutants may arise, e.g., all thermal cutting methods release aerosols, and hydromechanical methods produce fine particles of swarf. Similarly, radioactive liquids can remain in the equipment that is to be dismantled, lubricant fluids can leak from it, and waste solutions can be produced by decontamination procedures, air ventilation, and drains.

The decommissioning design for LLW consists of processes including:

- waste sorting, e.g., activated or not activated, combustible or not, steel type;
- segmentation and decontamination using various mechanical, thermal, and chemical methods;
- volume reduction by press compaction, incineration, evaporation, and melting;
- conditioning by embedding the waste in a cement or epoxy resin matrix, or by grouting it with cement;
- packaging; and,
- monitoring and tracking for intermediary storage [70].

Incineration is effective in reducing the volume of some solid waste like spent gloves, clothing, packing material, removed paint, and resins from water treatment. It is also used to reduce the volume of liquids containing organic compounds or using organic binders. Evaporation is suitable for reducing the volume of most aqueous liquid waste. Melting is specifically applied to reduce metal structures to ingots that later can be diluted with other metals for reuse or conveniently disposed of in near-surface repositories.

20% of reactors use graphite as a moderator. Among them are the gas-cooled reactor (GCR) and the advanced gas reactor (AGR) from the UK,

---

[13] It is a stainless-steel cylinder surrounding the nuclear reactor core.

the high-power channel reactor (RBMK) from Russia, and some innovative reactors such as the high-temperature gas-cooled reactor (HTGR), and the experimental pebble bed modular reactor. Many of the older reactors are now shut down and, worldwide, there is an accumulation of more than 250,000 tons of irradiated graphite that needs to be disposed of. Because graphite presents the highest induced activity and is also often contaminated with fission products, it is generally classified as ILW or long-lived LLW and means a real and unique challenge for decommissioning. Current deep disposal of the accumulated amount is impracticable, and graphite porosity deems it not suitable for shallow disposal [74].

The most relevant graphite contaminants are ^{14}C, ^{36}Cl, and ^{3}H and despite these being beta emitters, their long T½, volatility, and mobility in geological media make them hard to manage. Also, graphite may hold small quantities of fission products and actinides such as ^{134}Cs, ^{137}Cs, ^{90}Sr, ^{238}Pu, ^{241}Pu, and ^{238}U.

Recent studies have explored the possibilities of the disposal of the $^{14}CO_2$ produced by graphite oxidation through its release into the atmosphere using a method known as carbon capture and storage,[14] after reducing its concentration [75]. Additionally, these studies explored treatments by incineration; leaching/washing; heat treatments; thermal decontamination; and graphite immobilization in borosilicate glass, cement, and concrete. Also, graphite packaging and storage including *in-situ* entombment; *on-site* interim storage; surface disposal, shallow disposal, and deep disposal [76].

Consideration of future waste from advanced reactors

The management of future waste from advanced reactors is also influenced by the policy governing spent fuel reprocessing and/or disposal. As was learned earlier, two different paths have been identified: the open and the closed cycle. That is, reprocess and recycle the nuclear fuel to reduce the amount of high-activity, long-lived radioactive waste that needs very long-term storage, or dispose of the majority of spent fuel in deep geological repositories and leave future generations the task of managing the site. Neither of the two alternatives is completely resolved. A few countries[15] have accumulated wide experience in spent fuel recycling and

---

[14] A process to reduce the CO_2 from conventional industrial sources by separating, treating, and transporting it to a long-term storage location.

[15] France, the United Kingdom, and Russia are the countries with more experience in reprocessing spent nuclear fuel. India has limited capacity plants for reprocessing, while Japan has been constructing a reprocessing plant for more than 30 years now, and China is just beginning with its reprocessing plans.

reprocessing and only have to deal with a fraction of HLW for disposal in deep repositories. Just a few others, e.g., Finland, are close to operating their first repository.

While investing in expensive recycling technologies may not be a viable solution for most countries with nuclear power plants, finding a site for a repository with suitable geochemical conditions and a geologic environment can also be challenging. These complex issues may require global and cooperative solutions. For instance, the Orano Center in La Hague, France, has processed spent fuel from several countries including Germany, Switzerland, Belgium, Italy, the Netherlands, China, Japan, and Australia. It has also provided technology for current recycling projects in Japan and China. In existing partnerships, the waste resulting from reprocessing is returned to the country of origin for further storage and disposal [28].

The US generates around 2000 metric tons a year and, so far, the decision made has been their interim storage at more than 70 reactor sites across 35 states in wait for a geological repository [77]. Worldwide, more than 400,000 metric tons of spent fuel have been generated but only one-third of them are reprocessed. In a 2016 estimation, around 70% of spent nuclear fuel was wet-stored, while 30% was dry-stored [33].

Advanced reactors are being designed with diverse improved fuel types—TRISO, extruded metal fuel, liquid fuel, etc.—sharing MOX in higher rates; and higher levels of enrichment ($\geq 10\%$) and burnups (75–80 GWd/t). Some of them are fast reactors, and some also introduce the potential of fuel recycling [78]. These new reactors, which represent a huge development for clean power for the future, also bring with them new challenges for the management of radioactive waste whether the spent fuel and HLW are reprocessed or stored pending deep geological disposal.

Although some advanced reactor producers have announced benefits in the amount and harmful characteristics of radioactive waste from the new generation of nuclear reactors, there are known challenges in the composition and volume of waste and management approaches that require consideration [79].

The following is a brief review of some of their advantages and potential waste challenges. For example, innovative water-cooled reactors like SCWR and the AHWR foresee the introduction of thorium fuel to reduce the inventory of long-lived fission products. Gas-cooled reactors could produce less HLW thanks to the higher depletion of the spent fuel, and the absence of activated metals, however, graphite could be a problematic waste. Liquid-metal-cooled reactors involve advanced fuel reprocessing of transuranic elements with the potential to reduce waste generation, while the chemical toxicity of lead and bismuth could be a concern for long-term management. Molten sodium fast reactors are known to be actinide burners and, in molten salt reactors, the fertile and fissile nuclides are dissolved in the coolant and could support the regeneration of ^{233}U when transitioning from the common U-Pu cycle to the Th-U cycle, however, a solution is needed for the handling of the salt and its separation from fission products. Besides, waste forms and packages should be developed because of the corrosive nature of fluoride salts [80].

Moreover, advanced reactors will use different fuels. Given that in an open nuclear cycle spent fuel will always be a waste, there will be new streams of spent fuel containing not only uranium oxides, but uranium carbide, uranium nitride, metal uranium, and uranium molten salt. In a close nuclear cycle, wastes from nuclear fuel reprocessing will also differ from those of current reactors.

Sodium-cooled, gas-cooled, lead-cooled, and molten-salt-cooled reactors will generate during operation unique low and intermediate-level waste (LILW) needing long-term storage, which represents new challenges. For example, sodium-wetted metallic waste and fuel; lead-wetted metallic waste and fuel; ceramic waste (graphite prism and TRISO pebbles); activated or fuel-contaminated metallic and carbide waste; salt waste from cleaning operations, and bulk salts from the cooling system.

Various processes have already been proven in reactors such as Phenix, EBR-II, etc. from the 1960s–1990s to treat bulk sodium, residual sodium, and sodium-containing mixed waste avoiding the risk of violent reactions from sodium-cooled reactors. To reduce the radioactive contamination of Cs in the liquid sodium—the main gamma contaminant in the liquid phase—a method based on the use of traps loaded with a rigid carbon foam with a honeycomb structure called Reticulated Vitreous Carbon material (RVC) has been proven successful. To extract tritium from sodium and to eliminate sodium hydrides, cold trap systems have been employed. Bulk sodium has also been converted into sodium hydroxide by injecting small amounts of liquid sodium into a large flow of aqueous sodium hydroxide, circulating in a closed vessel at 60°C. Another process considered has been transforming sodium into a sodium carbonate form with carbon dioxide. Residual sodium has been removed using a vapor and nitrogen process resulting in sodium hydroxide and hydrogen generation [81].

Lead-fast reactors, which are anticipated to play an important role in the transmutation of actinides, could thereby reduce the volume of HLW from nuclear reactors. They can generate waste in the form of lead-wetted combustible, lead-wetted metallic waste, and bulk lead that must be treated by applying the restrictions and methods for handling lead and bismuth-contaminated materials. Likewise, in these processes, it is imperative to carefully control the temperature and prevent the formation of aerosols [80].

Activated or fuel-contaminated waste is produced in fast molten-salt-cooled reactors with liquid fuel. This waste can be removed by washing it in a clean fluoride bath [80]. Hydrogen fluoride that might be produced by fluoride hydrolysis during the process, could be removed using scrubbing methods, e.g., molten hydroxide scrubbers and solid sorbents [82].

Similarly, thermal molten-salt-cooled reactors, very high-temperature reactors, and gas-cooled reactors, while lowering radioactive releases because of the retention properties of the graphite and the helium coolant, could generate a great amount of activated graphite waste, e.g., skeletons of fuel assemblies, and spent fuel blocks, containing large amounts of ^{14}C and ^{3}H.

At this point, it is crucial to remember that the responsibility of waste management should not be passed on to future generations. Therefore, advanced reactor waste management solutions should not overlook the experience of multiple alternatives for minimizing and concentrating radioactive waste, including nuclear transmutation. It should also recognize the need to develop storage and transport containers for both new and irradiated high-assay low-enriched uranium (HALEU) fuels.

References

[1] United Nations. 2023. Climate Action. United Nations, n.d. [Online]. Available: https://www.un.org/en/climatechange/science/causes-effects-climate-change. [Accessed 20 August 2023].

[2] Energy Institute (EI). 2023. 72nd Statistical Review of World Energy. Energy Institute, London.

[3] Ritchie, H. and M. Roser. 2022. Energy mix. Our World in Data, n.d 2022. [Online]. Available: https://ourworldindata.org/electricity-mix. [Accessed 16 August 2023].

[4] Zeniewski, P., C. McGlade, T.-Y. Kim, T. Gould et al. 2019. The Role of Gas in Today's Energy Transitions. IEA, London.

[5] Lee, H. and J. Romero (eds.). 2023. IPCC, 2023: Sections. Climate Change 2023: Synthesis Report. Contribution of Working Groups I, II and III to the Sixth Assessment Report of the Intergovernmental Panel on Climate Change. IPCC, Geneva.

[6] Friedman, D. and J. Kotek. 2019. Changing the Game by Linking Nuclear and Renewable Energy Systems. Office of Energy Efficiency & Renewable Energy, 8 December 2019. [Online]. Available: https://www.energy.gov/eere/articles/changing-game-linking-nuclear-and-renewable-energy-systems. [Accessed 20 August 2023].

[7] Energy.gov. 2023. 5 Fast Facts About Nuclear Energy. Office of Nuclear Energy, US Department of Energy, April 2023. [Online]. Available: https://www.energy.gov/ne/articles/5-fast-facts-about-nuclear-energy. [Accessed 20 August 2023].

[8] World Nuclear Association. 2020. Naturally-Occurring Radioactive Materials (NORM). World Nuclear Association, April 2020. [Online]. Available: https://world-nuclear.org/information-library/safety-and-security/radiation-and-health/naturally-occurring-radioactive-materials-norm.aspx. [Accessed 20 August 2023].

[9] Nuclear Energy Institute (NEI). 2015. Safety: The Nuclear Energy Industry's Highest Priority. Nuclear Energy Institute, June 2015. [Online]. Available: https://www.nei.org/resources/fact-sheets/safety-nuclear-energy-industry-highest-priority. [Accessed 27 October 2023].

[10] International Atomic Energy Agency. 2023. Now Available: Upgraded IAEA Database on Discharges of Radionuclides from Nuclear Installations. IAEA, 1 August 2023. [Online]. Available: https://www.iaea.org/newscenter/news/now-available-upgraded-iaea-database-on-discharges-of-radionuclides-from-nuclear-installations. [Accessed 21 August 2023].

[11] International Atomic Energy Agency. 2023. Database on Discharges of Radionuclides to the Atmosphere and Aquatic Environment. IAEA, n.d. [Online]. Available: https://www.iaea.org/resources/databases/dirata. [Accessed 27 October 2023].

[12] International Atomic Energy Agency. 2023. Joint Convention on the Safety of Spent Fuel Management and on the Safety of Radioactive Waste Management. IAEA, n.d. [Online]. Available: https://www.iaea.org/topics/nuclear-safety-conventions/joint-convention-safety-spent-fuel-management-and-safety-radioactive-waste. [Accessed 21 August 2023].

[13] Domenech, H. 2017. Radiation Safety. Management and Programs, Switzerland: Springer International.

[14] International Atomic Energy Agency. 2021. Safety Series No. SSG-60. Management of Residues Containing Naturally Occurring Radioactive Material from Uranium Production and Other Activities, Vienna: IAEA.

[15] Federal Government. 2023. Title 10 Chapter I Part 40 Appendix A to Part 40—Criteria Relating to the Operation of Uranium Mills and the Disposition of Tailings or Wastes... National Archives Code of Federal Regulations, 14 August 2023. [Online]. Available: https://www.ecfr.gov/current/title-10/chapter-I/part-40/appendix-Appendix%20A%20to%20Part%2040. [Accessed 22 August 2023].

[16] Diehl, P. 2011. Uranium Mining and Milling Wastes: An Introduction. World Information Service on Energy (WISE), 18 May 2011. [Online]. Available: http://wise-uranium.org/uwai.html#WASTE. [Accessed 22 August 2023].

[17] US Department of Energy. 2018. Legacy Management UMTRACA Titles I & II Fact Sheet, Washington: DOE.

[18] Energy.gov. 2023. Moab By the Numbers. Office of Environmental Management Department of Energy, June 2023. [Online]. Available: https://www.energy.gov/em/moab-numbers. [Accessed 23 August 2023].

[19] Energy.gov. 2023. Crescent Junction Site Operations. Office of Environmental Management Department of Energy, n.d. [Online]. Available: https://www.energy.gov/em/moab/crescent-junction-site-operations. [Accessed 22 August 2023].

[20] Office of Environmental Management US Department of Energy. 2023. DOE-EM/GJRAC3102 Moab UMTRA Project 2022 Groundwater Program Report. DOE, Washington.

[21] US NRC. 2020. Low-Level Waste. Nuclear Regulatory Commission, 12 March 2020. [Online]. Available: https://www.nrc.gov/waste/low-level-waste.html. [Accessed 28 August 2023].

[22] World Nuclear Association. 2022. Conversion and Deconversion. World Nuclear Association, January 2022. [Online]. Available: https://world-nuclear.org/information-library/nuclear-fuel-cycle/conversion-enrichment-and-fabrication/conversion-and-deconversion.aspx. [Accessed 29 August 2023].

[23] US DOE Office of Environmental Management. 2019. Nuclear Regulatory Commission's Low-Level Radioactive Waste Classifications, Washington: Department of Energy.

[24] International Atomic Energy Agency. 2023. Depleted Uranium. IAEA, n.d. [Online]. Available: https://www.iaea.org/topics/spent-fuel-management/depleted-uranium. [Accessed 31 August 2023].

[25] Orano. 2022. International expert in uranium processing. Orano Group [Online]. Available: https://www.orano.group/en/nuclear-expertise/from-exploration-to-recycling/international-expert-in-uranium-processing. [Accessed 13 November 2022].

[26] Urenco ChemPlants. 2023. Tails Management Facility. Urenco The Energy to Succeed, n.d 2023. [Online]. Available: https://www.urenco.com/global-operations/urenco-chemplants. [Accessed 1 September 2023].

[27] Orano. 2023. Everything you need to know about France's National plan for the management of radioactive materials and waste. Orano, n.d. [Online]. Available: https://www.orano.group/en/unpacking-nuclear/everything-you-need-to-know-about-france-s-national-plan-for-the-management-of-radioactive-materials-and-waste. [Accessed 29 August 2023].

[28] Orano. 2023. The world leader in recycling used nuclear fuels. Orano, n.d 2023. [Online]. Available: https://www.orano.group/en/nuclear-expertise/from-exploration-to-recycling/world-leader-in-recycling-used-nuclear-fuels. [Accessed 1 September 2023].

[29] National Academy of Sciences. 2023. Merits, and Viability of Different Nuclear Fuel Cycles and Technology Options and the Waste Aspects of Advanced Nuclear Reactors, Washington: National Academies Press.

[30] International Atomic Energy Agency. 1996. IAEA-TECDOC-918 Health and Environmental Aspects of Nuclear Fuel Cycle Facilities, Vienna: IAEA.

[31] Nuclear Decommissioning Authority (NDA). FACTSHEET: Spent Fuel Reprocessing, London: NDA, n.d.

[32] Le Bar, S. 1996. Environmental safety of reprocessing plant La Hague. pp. 105–112. In IAEA-TECDOC-918 Health and Environmental Aspects of Nuclear Fuel Cycle Facilities, Vienna, IAEA.

[33] International Atomic Energy Agency. 2022. Nuclear Energy Series No. NW-T-1.14 (Rev. 1). Status and Trends in Spent Fuel and Radioactive Waste Management, Vienna: IAEA.

[34] Londres, V., R. Do Quang and P. Fournier. 2003. Management of waste from french nuclear fuel cycle: what are the key issues? In International Conference on the Safety of Radioactive Waste Management, Cordoba.

[35] ANDRA International. 2023. Project siting and facilities overview. ANDRA International, n.d. [Online]. Available: https://international.andra.fr/projects/cigeo/cigeos-facilities-and-operation/project-siting-and-facilities-overview. [Accessed 7 September 2023].

[36] Dalton, D. 2022. Sweden/Government Approves SKB's Plans For Final Repository At Forsmark For Spent Nuclear Fuel. NUCNET (Nuclear News Agency), 28 January 2022. [Online]. Available: https://www.nucnet.org/news/government-approves-skb-s-plans-for-final-repository-for-spent-nuclear-fuel-1-5-2022. [Accessed 7 September 2023].

[37] Waste Isolation Pilot Plant (WIPP). 2023. WIPP SITE. US Department of Energy, Salado Isolation Mining Contractors LLC (SIMCO), n.d. [Online]. Available: https://wipp.energy.gov/wipp-site.asp. [Accessed 7 September 2023].

[38] Swedish Nuclear Fuel and Waste Management Company SKB. 2021. The Final Repository SFR. This is where Sweden keeps its radioactive operational waste. Svensk Kärnbränslehantering AB (SKB), 9 March 2021. [Online]. Available: https://skb.com/our-operations/sfr/. [Accessed 7 September 2023].

[39] Swedish Nuclear Fuel and Waste Management Company SKB. 2021. The Swedish System. A safe final repository system. Svensk Kärnbränslehantering AB (SKB),

4 February 2021. [Online]. Available: https://skb.com/our-operations/the-swedish-system/. [Accessed 7 September 2023].

[40] Posiva Oy. 2023. Final Disposal, Repository in ONKALO. Posiva Oy Olkiluoto, n.d. [Online]. Available: https://www.posiva.fi/en/index/finaldisposal/researchandfinaldisposalfacilitiesatonkalo.html. [Accessed 8 September 2023].

[41] Posiva Oy. 2023. Final disposal tests. Posiva Oy Olkiluoto, n.d. [Online]. Available: https://www.posiva.fi/en/index/finaldisposal/finaldisposaltestsfisstytk.html#. [Accessed 9 September 2023].

[42] US NRC. 2023. Radioactive Waste Management Technology Chapter 7 Gaseous Radioactive Wastes, Chattanooga: USNRC Technical Training Center.

[43] International Atomic Energy Agency. 2016. IAEA Safety Standards No. SSG-40. Predisposal Management of Radioactive Waste from Nuclear Power Plants and Research Reactors, Vienna: IAEA.

[44] US NRC. 2023. Radioactive Waste Management Technology Chapter 4 Liquid Radioactive Wastes, Chattanooga: USNRC Technical Training Center.

[45] US NRC. 2022. LLRW Toolbox: Decay-In-Storage (DIS). US Nuclear Regulatory Commission, 2 August 2022. [Online]. Available: https://www.nrc.gov/materials/toolboxes/llrw/decay-in-storage.html. [Accessed 10 September 2023].

[46] Federal Government. 2023. Title 10 Chapter I Part 20 Standards for Protection Against Radiation. National Archives Code of Federal Regulations, 5 September 2023. [Online]. Available: https://www.ecfr.gov/current/title-10/chapter-I/part-20. [Accessed 10 September 2023].

[47] International Atomic Energy Agency. 2009. Safety Guide No. GSG-1 Classification of Radioactive Wastes, Vienna: IAEA.

[48] Federal Government. 2023. Title 10 Chapter I Part 61 Paragraph 61.55 Classification for near-surface disposal. National Archives Code of Federal Regulations, 7 September 2023. [Online]. Available: https://www.ecfr.gov/current/title-10/chapter-I/part-61/subpart-D/section-61.55. [Accessed 10 September 2023].

[49] National Academies of Sciences. 2006. Committee on Improving Practices for Regulating and Managing Low-Activity Radioactive Wastes, National Research Council, Improving the Regulation and Management of Low-Activity Radioactive Wastes, Washington: National Academies Press.

[50] French National Agency for Radioactive Waste Management (ANDRA). 2023. Operational Facilities. ANDRA, n.d. [Online]. Available: https://international.andra.fr/operational-facilities. [Accessed 13 September 2023].

[51] ENRESA. 2023. El Cabril Disposal Facility. ENRESA, n.d. [Online]. Available: https://www.enresa.es/eng/index/activities-and-projects/el-cabril. [Accessed 13 September 2023].

[52] Swedish Nuclear Fuel and Waste Management Company SKB. SFR – Final Repository for Short-lived Radioactive Waste Brochure, Östhammar: Svensk Kärnbränslehantering AB (SKB), n.d.

[53] GOV.UK. 2022. Corporate report UK Radioactive Waste Inventory 2022. Department for Business, Energy & Industrial Strategy and Nuclear Decommissioning Authority, 8 February 2023. [Online]. Available: https://www.gov.uk/government/publications/uk-radioactive-waste-and-material-inventory-2022/uk-radioactive-waste-inventory-2022#volumes-by-region. [Accessed 13 September 2023].

[54] US NRC. 2020. Locations of Low-Level Waste Disposal Facilities. Nuclear Regulatory Commission, 12 March 2020. [Online]. Available: https://www.nrc.gov/waste/llw-disposal/licensing/locations.html. [Accessed 14 September 2023].

[55] US NRC. 2023. NUREG-1350 Vol. 34 Section 4 Radioactive Waste. Nuclear Regulatory Commission, Washington.

[56] Waste Control Specialists (WCS). 2023. Facilities. Waste Control Specialists (WCS), n.d. [Online]. Available: https://www.wcstexas.com/about/our-facilities/facilities/. [Accessed 14 September 2023].

[57] South Carolina Department of Health and Environmental Control. Commercial Low-Level Radioactive Waste Disposal In South Carolina Booklet, Columbia: South Carolina Department of Health and Environmental Control, CR-000907 3/07.

[58] Energy Solutions. 2023. Clive Disposal Facility. Energy Solutions, n.d. [Online]. Available: https://www.energysolutions.com/our_facilities/clive-disposal-facility/. [Accessed 15 September 2023].

[59] Utah Division of Waste Management and Radiation Control. 2023. Low-Level Radioactive Waste: Energy Solutions. Utah Department of Environmental Quality, 2023. [Online]. Available: https://deq.utah.gov/businesses-facilities/low-level-radioactive-waste-energysolutions. [Accessed 15 September 2023].

[60] International Atomic Energy Agency. 2020. IAEA Nuclear Energy Series No. NW-T-1.27 Design Principles and Approaches for Radioactive Waste Repositories, Vienna: IAEA.

[61] Gov.UK. 2023. Sellafield makes history with first retrievals from oldest store. Sellafield Ltd and Nuclear Decommissioning Authority, 16 August 2023. [Online]. Available: https://www.gov.uk/government/news/sellafield-makes-history-with-first-retrievals-from-oldest-store. [Accessed 13 September 2023].

[62] The Henry L. Stimson Center. 2019. Exploring the Wolsung LILW Disposal Center in South Korea. The Henry L. Stimson Center, 7 August 2019. [Online]. Available: https://www.stimson.org/2019/exploring-wolsung-lilw-disposal-center-south-korea/. [Accessed 16 September 2023].

[63] World Nuclear News. 2022. Expansion of South Korean waste repository begins. World Nuclear Association, 31 August 2022. [Online]. Available: https://www.world-nuclear-news.org/Articles/Expansion-of-South-Korean-waste-repository-begins. [Accessed 16 September 2023].

[64] Agence Nationale pour la Gestion des Déchets Radioactifs (ANDRA). 2023. Safety: anticipating the risks. French National Radioactive Waste Management Agency, n.d. [Online]. Available: https://international.andra.fr/projects/cigeo/safety-anticipating-risks. [Accessed 18 September 2023].

[65] French National Radioactive Waste Management Agency (ANDRA). 2020. The Cigeo Project. France's Industrial Centre for Geological Disposal of Radioactive Waste Brochure, Châtenay-Malabry: ANDRA.

[66] French National Radioactive Waste Management Agency (ANDRA). 2023. Submission of the application for authorization to create Cigéo. French National Radioactive Waste Management Agency (ANDRA), 17 January 2023. [Online]. Available: https://international.andra.fr/submission-application-authorization-create-cigeo. [Accessed 18 September 2023].

[67] US Environmental Protection Agency (EPA). 2022. What is the Yucca Mountain repository? US Environmental Protection Agency (EPA), 30 September 2022. [Online]. Available: https://www.epa.gov/radiation/what-yucca-mountain-repository. [Accessed 18 September 2023].

[68] Eckhardt, R. C. 2000. Yucca Mountain. Looking ten thousand years into the future. Los Alamos Science, vol. 26.

[69] Nuclear Energy Institute (NEI). 2019. Decommissioning Nuclear Power Plants. Nuclear Energy Institute, August 2019. [Online]. Available: https://nei.org/resources/fact-sheets/decommissioning-nuclear-power-plants. [Accessed 20 September 2023].

[70] Waste Management & Decommissioning Working Group. 2019. World Nuclear Association, Methodology to Manage Material and Waste from Nuclear Decommissioning. World Nuclear Association, London.

[71] US NRC. 2021. Generic Environmental Impact Statement for License Renewal of Nuclear Plants (NUREG-1437 Vol. 1, Part 7). Nuclear Regulatory Commission, 24 March 2021. [Online]. Available: https://www.nrc.gov/reading-rm/doc-collections/nuregs/staff/sr1437/v1/part07.html#1%20Nuclear%20Power%20Plants. [Accessed 20 September 2023].

[72] Evans, J. C., E. L. Lepel, R. W. Sanders, C. L. Wilkerson, W. Silker, C. W. Thomas et al. 1984. NUREG/CR-3474 Long-Lived Activation Products in Reactor Materials. Pacific Northwest Laboratory, Richland.

[73] OECD Nuclear Energy Agency (NEA). 2014. R&D and Innovation Needs for Decommissioning Nuclear Facilities. Organization for Economic Cooperation and Development (OECD), Paris.

[74] Wickham, A., H.-J. Steinmetz, P. O'Sullivan and M. I. Ojovan. 2017. Updating irradiated graphite disposal: Project 'GRAPA' and the international decommissioning network. Journal of Environmental Radioactivity 171: 34–40.

[75] Wickham, A., M. Ojovan and P. O'Sullivan. 2017. Innovative approaches to the management of irradiated nuclear graphite wastes: addressing the challenges through international collaboration with Project 'GRAPA'. In Workshop on Current and Emerging Methods for Optimizing Safety and Efficiency in Nuclear Decommissioning, Sarpsborg.

[76] Ojovan, M. I. and A. J. Wickham. 2017. Processing of Irradiated Graphite: The Outcomes of an IAEA Coordinated Research Project. MRS Advances 1(62): 1–6.

[77] Energy.gov. 2022. 5 Fast Facts about Spent Nuclear Fuel. Office of Nuclear Energy, Department of Energy, 3 October 2022. [Online]. Available: https://www.energy.gov/ne/articles/5-fast-facts-about-spent-nuclear-fuel. [Accessed 26 September 2023].

[78] Oklo Inc. 2023. Oklo Prepares for the Deployment of its Commercial-Scale Fuel Recycling Facility with the Submission of its Licensing Project Plan. Oklo Inc., 25 January 2023. [Online]. Available: https://oklo.com/newsroom/news-details/2023/Oklo-Prepares-for-the-Deployment-of-its-Commercial-Scale-Fuel-Recycling-Facility-with-the-Submission-of-its-Licensing-Project-Plan/default.aspx. [Accessed 27 September 2023].

[79] Price, R. 2021. Bringing the Back-End to the Forefront. Spent Fuel Management and Safeguards Considerations for Emerging Reactors. The Henry L. Stimson Center, 10 February 2021. [Online]. Available: https://www.stimson.org/2021/bringing-the-back-end-to-the-forefront/. [Accessed 29 September 2023].

[80] International Atomic Energy Agency. 2019. IAEA Nuclear Energy Series No. NW-T-1.7. Waste from Innovative Types of Reactors and Fuel Cycles. A Preliminary Study. IAEA, Vienna.

[81] International Atomic Energy Agency. 2007. IAEA-TECDOC-1534 Radioactive Sodium Waste Treatment and Conditioning. Review of Main Aspects, Vienna: IAEA.

[82] Riley, B. J., J. McFarlane, G. D. DelCul, J. D. Vienna, C. I. Contescu and C. W. Forsberg. 2019. Molten salt reactor waste and effluent management strategies: A review. Nuclear Engineering and Design 345(15): 94–109.

²³⁵UF₆ cylinder and over-pack for
transport 61

A

accident-tolerant fuels 112–116, 118,
127, 158
acid and alkaline leaching processes 49,
51, 166
ammonium diuranate (ADU) 51, 53
analysis of postulated events 89, 152, 154
ANEEL fuel technology 119
authorization requirements 30, 61, 82, 130,
142, 144

B

bituminized waste 173
blast hole uranium mining 46
BN-600 reactor 68, 109, 110, 133
BN-800 reactor 68, 109, 110, 133
boiling water reactor (BWR) 20, 22, 36, 60,
97, 101
boxhole uranium mining 46, 47

C

CANDU reactor 99, 102, 103, 105, 106
cemented waste 173, 179, 186
central interim storage facilities 67
centrifuge uranium enrichment 51, 55–57
chain reaction 8, 9, 12, 13, 22, 81, 90, 107,
142, 145, 157
chemical hazards 87, 89, 171
CLAB 67, 174
clearance level 176, 177, 185
Colorado River 167–169
commercial nuclear reactors 30, 60, 153,
156, 165

concentrate and contain 52
containment 23, 28, 35–37, 71, 86, 90,
104–108, 116, 117, 121, 124, 128, 131,
143, 147, 148, 150–152, 158, 159, 178,
180, 184
conventional open pit uranium mine 43
conventional underground uranium
mine 45
conversion into UF₆ 52
core catcher 110, 121–124, 158
CRISLA process 59
critical excursion 90
criticality 62, 65, 81, 88–92, 111, 133,
144, 145

D

decommissioning 35, 66, 67, 82, 111, 142,
144, 146, 180, 182, 184–187
Deconversion 42, 70–73, 171, 172
deep geological disposal 183, 188
deep geological repository 42, 67
defense-in-depth 87, 149–152, 157
defense-in-depth approach 87
delay and decay 176
depleted uranium 33, 42, 58, 70, 72, 73, 85,
111, 118, 170–172
depleted uranium hexafluoride (DUF6)
deconversion 42, 70, 171
design-basis and non-design-basis accidents
113, 148, 149
dilute and disperse 177
Discharges of Radionuclides to the
Atmosphere and Aquatic Environment
(DIRATA) 165, 166
disposal cell 167–169
dose limits 86, 143, 165, 166
dry casks for spent fuel storage 42, 65–67

198 *Nuclear Materials: An Overview*

dry conversion process 52, 82
dry spent fuel interim storage 64, 65, 67, 68,
     86, 174, 182, 184
DUF6 42, 58, 70–74, 171

E

emergency planning 93, 153, 154, 158
emergency response preparedness 155
enriched UO_2 60, 100
evolutionary reactors 121
Experimental Breeder Reactor
     (EBR-I) 18, 108
Experimental Breeder Reactor
     (EBR-II) 108, 189
exposure to gamma radiation 87
exposure to radon hazard 45, 46
extruded metal fuel 113, 188

F

Facilities 14, 23, 24, 27, 29–32, 34, 35, 45,
     48, 50, 55, 60, 67, 72, 81, 82, 84, 86, 89,
     93, 112, 118, 146, 154, 170–172, 174,
     178–181, 183, 184, 186
fast breeder reactor (FBR) 112
first post-war nuclear reactors 18
French fast reactor 22, 110, 134
fuel assemblies 32, 36, 37, 63–65, 68, 82, 99,
     100, 102, 107, 111, 124, 176, 190
fuel burnup 64, 172
fuel cladding 113, 115, 148, 175, 182
fuel development 112
fuel pellets 32, 62, 64, 69, 107, 113, 118,
     148, 174
Fuel-in-Fabric (F-in-F) fuel 118

G

gas-cooled reactor (GCR) 60, 99, 102, 120,
     126, 127, 133, 157, 186–188, 190
gaseous diffusion uranium enrichment 14,
     55, 56
gaseous effluents 170
generation IV reactors 119, 134
global contribution by energy source 164
greenhouse gas emissions 164
GTCC (greater than C) waste 171

H

high assay low-enriched uranium 112, 119,
     127, 190
high level radioactive waste (HLW) 67, 68,
     116, 182
high-level waste (HLW) 68, 178, 179
high-temperature gas reactor (HTGCR) 60,
     99, 126, 127, 129, 133, 157, 187
hydrogen recombiners 122, 158

I

IAEA inspectors 29, 31
IAEA Safeguards and Verification System
     25, 29
Impoundment 88, 167, 169
increased reliability of advanced
     reactors 157
India fast reactor 112
Indian nuclear energy program 111
information system 32, 68, 120, 133
innovative advanced cladding 113
in-situ uranium leaching 48, 82, 85, 87
intermediate-level waste (ILW) 178
internal exposure 82, 83, 86
irradiated graphite 187

J

Japanese fast reactor 111
jet boring uranium mining 47

K

key measurement points 32, 33

L

laser uranium isotope separation 58, 59
light water graphite moderated reactor
     (LWGR) 19, 60
liquid effluents 88, 171, 173
liquid fuels 118, 129, 188, 190
liquid waste treatment 170
LLW waste from uranium conversion 52,
     72, 170
locations outside facilities (LOF) 29
long-lived 83, 170, 171, 173, 174, 177, 178,
     182, 185, 187, 188

long-term storage 70, 72, 73, 116, 118, 173,
179, 183, 187, 189
loop-type design 109
loss-of-coolant accident (LOCA) 99
low level waste (LLW) from mining and
milling 170
low-enriched uranium fuel fabrication 62
low-level waste (LLW) 171, 174, 178,
180, 181

M

management of radioactive waste 82, 86,
88, 142, 171, 176, 177, 180, 188
material balance areas 31
Member States non proliferation legal
framework 23
Member States responsibilities 25, 34
Microreactors 129–132, 134
mill tailings remedial actions 167
minimization of radioactive waste 176
minimization of waste 176
minor actinides 68–70, 110, 134, 172
mitigation measures 90, 92
mixed (MOX) fuel 42, 68, 100, 110–112
Moab 167–169

N

natural radioactive element 165
natural UO$_2$ 60, 61, 100, 102, 103
near-surface disposal 174, 178–180
neutron discovery 11
new fuel pellet designs 113
non proliferation and nuclear energy 23, 25
non-entry mining methods 46
Non-Proliferation Treaty (NPT) 25, 28
nuclear facility security 81
nuclear fission 7, 10–12, 14, 25, 81
Nuclear fuel cycle stages 42, 82, 170
nuclear fuel reprocessing 172, 189
nuclear fusion 9, 24
nuclear power 19–23, 37, 57, 60, 67, 69,
96–98, 108–111, 113, 115, 119, 125, 130,
131, 146, 154, 164–167, 172, 180,
184, 188
nuclear reactions 6, 8, 11, 12, 14
Nuclear reactor 4, 6, 8, 13, 14, 18–20, 30, 32,
33, 42, 60, 68, 96–100, 104, 108, 112, 119,
121, 134, 142, 145–147, 149, 153, 154,
156, 165, 175, 179, 184, 186, 188, 189

nuclear safety 81, 88, 128, 142, 144, 146, 151,
156, 184
nuclear weapon tests 22, 23

O

off-gas 54, 69, 88, 172, 175, 176
Onkalo 67, 174, 175
operational occurrences 148, 150, 151
ore milling 43, 83

P

partitioning of minor actinides 70
passive heat pipe-cooling 129
passive safety systems 121, 125, 127,
134, 158
physical protection 30, 37, 146, 147
pool-type design 108, 109, 111, 112
pressurized heavy water reactor (PHWR)
60, 119
pressurized water reactor (PWR) 36, 60,
97, 100
prevention of accidents 142, 144
preventive measures 90
Production of nuclear energy 13
PUREX process 69, 70

R

radiation protection program 83, 143
radiation protection requirements 88
Radiation risks 81, 142
radioactive decay 2, 3, 6, 45, 67, 89
radioactive material 1, 3, 5, 15, 24, 30, 48,
51, 52, 54, 81, 82, 84, 86, 87, 89, 90, 93,
104, 117, 142–144, 146–148, 150–152,
155, 157, 166, 167
radioactive nuclide 1
radioactive series 10
radioactive solid waste from mining 172
radioactive waste generation 134, 165
Radioactivity 1, 3, 5, 6, 11, 12, 88, 105, 106,
166, 170, 171, 174, 178, 179, 182, 184
Radionuclide 1, 3, 5–7, 10, 11, 14, 25, 45,
70, 83, 153, 158, 165, 166, 170, 171, 175,
177–179, 182, 184
radon air contamination 45, 46, 82–84,
87, 88
radon water contamination 88
raise bore uranium mining 46, 47
RBMK reactor 101, 107

reactor radiation safety 45, 81, 82, 142,
    143, 156
reactor vessel 98, 99, 103, 109, 110, 124, 126,
    128, 151, 158, 159, 185
reactors in operation 23, 98, 99
reactors recently connected to the grid
    20, 132
redundancy, diversity, and
    independence 151
re-enrichment 42, 70, 74
Renewable energy 164
Retrieval 179, 182, 183

S

safeguard measures 93
safeguards agreements 23, 26, 27, 29, 30,
    34, 38, 112
safety culture 151, 156, 157
Safety measures 45, 90, 93, 142
safety of facilities 81
safety risk assessment 144
safety systems, structures, and components
    147, 148, 151
security and safety interface 92, 93
segregation of radioactive waste 171
SILEX process 58, 59, 74
Small Modular Reactors (SMR) 120,
    121, 146
sodium diuranate (SDU) 51
sodium-cooled reactor 109, 110, 133, 189
source nuclear material 25
special nuclear material 25, 33, 70
spent fuel disposal 70
spent fuel interim storage 64, 68, 86, 182
spent fuel reprocessing 67, 87, 112, 172, 187
state-level approach 27
surveillance and containment systems 28
Swedish multi-barrier principle 174
System of Accounting for and Control of
    Nuclear Material (SSAC) 28
system, structures and components
    important to safety 113, 144, 146–151

T

tails of depleted uranium hexafluoride 171
technological waste 173
the atomic bomb 12, 13, 22

the International Atomic Energy Agency
    (IAEA) 22–38, 61, 81, 112, 120, 133,
    147–149, 155, 165
thermal reactors 99, 100, 103, 108, 126, 128
time-distance-shielding 85
total nuclear energy capacity 98
Transient Reactor test Facility (TREAT)
    115, 116
transmutation of minor actinides 70, 134
treatment of radioactive waste 8
tri-structural isotropic (TRISO) fuel
    116, 117

U

U_3O_8 package 51
UF6 cylinders for storage and transport
    61, 62
UK fast reactor 22, 111
UO_2 natural fuel fabrication 60–62
uranium concentrate (yellowcake)
    production 48, 52, 83
uranium conventional lixiviants 48
uranium enrichment 22, 34, 55, 60, 64, 70,
    87, 118
uranium mill tailings 166, 167
uranium mining 43, 86, 87
uranium precipitation 49
uranium solvent extraction 48–50, 52, 53, 69
uranium tails 58, 72, 88

V

very-low-level waste (VLLW) 173
vitrification 69, 172, 173

W

waste class 178
waste conditioning 82
waste facilities 171, 181
waste from advanced reactors 187
waste rock 45–48, 85, 166, 167
wet conversion process 82
wet spent fuel interim storage 68, 182, 184